T0327637

Multi-voltage CMOS Circuit Design

Multi-voltage CMOS
Circuit Design

Volkan Kursun
University of Wisconsin-Madison, USA

Eby G. Friedman
University of Rochester, USA

John Wiley & Sons, Ltd

Other Wiley Editorial Offices

John Wiley & Sons Inc., 111 River Street, Hoboken, NJ 07030, USA

Jossey-Bass, 989 Market Street, San Francisco, CA 94103-1741, USA

Wiley-VCH Verlag GmbH, Boschstr. 12, D-69469 Weinheim, Germany

John Wiley & Sons Australia Ltd, 42 McDougall Street, Milton, Queensland 4064, Australia

John Wiley & Sons (Asia) Pte Ltd, 2 Clementi Loop #02-01, Jin Xing Distripark, Singapore 129809

John Wiley & Sons Canada Ltd, 6045 Freemont Blvd, Mississauga, ONT, L5R 4J3

Wiley also publishes its books in a variety of electronic formats. Some content that appears in print may not be
available in electronic books.

Library of Congress Cataloging-in-Publication Data

Kursun, Volkan.
  Multi-Voltage CMOS Circuit Design / Volkan Kursun, Eby G. Friedman.
     p. cm.
  Includes bibliographical references and index.
  ISBN-13: 978-0-470-01023-5 (cloth : alk. paper)
  ISBN-10: 0-470-01023-1 (cloth : alk. paper)
  1.   Metal oxide semiconductors, Complementary.   I.   Friedman, Eby G.   II.
Title.
  TK7871.99.M44K87 2006
  621.39′732–dc22

                                                                    2006006472

British Library Cataloguing in Publication Data

A catalogue record for this book is available from the British Library

ISBN-13 978-0-470-01023-5
ISBN-10 0-470-01023-1

Typeset in 10/12 pt Times by Thomson Digital.

This book is printed on acid-free paper responsibly manufactured from sustainable forestry
in which at least two trees are planted for each one used for paper production.

This book is dedicated to the memory of my grandparents
Gülizar and Bahri

To the next generation
Joe, Samuel, Jesse, Jake, Hanan, and Josh

MELIORA

Contents

About the Authors

Volkan Kursun was born in Ankara, Turkey on June 5, 1974. He attended the Middle East Technical University from 1995 to 1999 and graduated with a Bachelor of Science degree in Electrical and Electronics Engineering in 1999. He attended the University of Rochester from 1999 to 2004 and received a Master of Science degree in Electrical and Computer Engineering in 2001 and a Doctor of Philosophy degree in Electrical Engineering in 2004.

He performed research on high speed voltage interface circuits with Xerox Corporation, Webster, New York in 2000. During the summers of 2001 and 2002, he was with Intel Microprocessor Research Laboratories, Hillsboro, Oregon, responsible for the modeling and design of high frequency monolithic DC–DC converters. He joined the Department of Electrical and Computer Engineering at the University of Wisconsin-Madison as an assistant professor in 2004.

His current research interests include low-voltage, low-power, and high-performance integrated circuit design, modeling of semiconductor devices, and emerging integrated circuit technologies. He has more than forty publications and two issued and four pending patents in the areas of high performance integrated circuits and emerging semiconductor technologies. He is a member of the technical program and organizing committees of a number of IEEE and ACM conferences. Dr. Kursun is a member of the editorial boards of the *IEEE Transactions on Circuits and Systems II* and the *Journal of Circuits, Systems and Computers*.

Eby G. Friedman received the B.S. degree from Lafayette College in 1979, and the M.S. and Ph.D. degrees from the University of California, Irvine, in 1981 and 1989, respectively, all in electrical engineering.

From 1979 to 1991, he was with Hughes Aircraft Company, rising to the position of manager of the Signal Processing Design and Test Department, responsible for the design and test of high-performance digital and analog IC's. He has been with the Department of Electrical and Computer Engineering at the University of Rochester since 1991, where he is a Distinguished Professor, the Director of the High Performance VLSI/IC Design and Analysis Laboratory, and the Director of the Center for Electronic Imaging Systems. He is also a Visiting Professor at the Technion - Israel Institute of Technology. His current research and teaching interests are in high performance synchronous digital and mixed-signal microelectronic design and analysis with application to high speed portable processors and low power wireless communications.

He is the author of more than 300 papers and book chapters, several patents and the author or editor of eight books in the fields of high speed and low power CMOS design techniques, high speed interconnect, and the theory and application of synchronous clock and power distribution networks. Dr. Friedman is the Regional Editor of the *Journal of Circuits, Systems and Computers*, a Member of the editorial boards of the *Analog Integrated Circuits and Signal Processing, Microelectronics Journal, Journal of Low Power Electronics* and *Journal of VLSI Signal Processing*, Chair of the *IEEE Transactions on Very Large Scale Integration (VLSI) Systems* steering committee, and a Member of the technical program committee of a number of conferences. He previously was the Editor-in-Chief of the *IEEE Transactions on Very Large Scale Integration (VLSI) Systems*, a Member of the editorial board of the *Proceedings of the IEEE* and *IEEE Transactions on Circuits and Systems II: Analog and Digital Signal Processing*, a Member of the Circuits and Systems (CAS) Society Board of Governors, CAS liaison to the Solid-State Circuits Society, Chair of the VLSI Systems and Applications CAS Technical Committee, Chair of the Electron Devices Chapter of the IEEE Rochester Section, Program and Technical chair of several IEEE conferences, Guest Editor of several special issues in a variety of journals, and a recipient of the Howard Hughes Masters and Doctoral Fellowships, an IBM University Research Award, an Outstanding IEEE Chapter Chairman Award, the University of Rochester Graduate Teaching Award, and a College of Engineering Teaching Excellence Award. Dr. Friedman is a Senior Fulbright Fellow and an IEEE Fellow.

Preface

The scaling of semiconductor process technologies has been continuing for more than four decades. Advancements in process technologies are the fuel that has been moving the semiconductor industry. In response to growing customer demand for applications that require integrated circuits with enhanced performance and functionality at reduced cost, a new process generation has been introduced by the semiconductor industry every two to three years during the past forty years. The challenge for enhancing the performance and functionality of integrated circuits has traditionally been managing the greater design and manufacturing complexity and higher power consumption. As stressed throughout this book, the generation, distribution, and dissipation of power are at the forefront of current problems faced by integrated circuit designers.

In order to continue the historical trend of reducing the unit cost of a circuit while simultaneously enhancing performance and functionality, radical changes are required in the manner in which integrated circuits are designed. Higher speed at all costs is no longer an option. Energy-efficient semiconductor devices, circuit techniques, and microarchitectures are necessary to maintain the pace of expansion that the semiconductor industry has enjoyed over the past forty years.

Several important opportunities that exist for low power and reliable integrated circuit and system design are highlighted in this book. Design choices that can be made while scaling the supply and threshold voltages, in order to lower power consumption and enhance device reliability without degrading circuit speed, are described. Techniques for simultaneously achieving energy efficiency and high speed are presented.

Systems with multiple power supplies can significantly reduce power consumption without degrading speed by selectively lowering the supply voltage along non-critical delay paths. High-frequency monolithic DC–DC conversion techniques applicable to multiple supply voltage CMOS circuits are presented that provide additional voltage levels with low energy and area overhead. Full integration of a high efficiency buck converter on the same die as a dual supply voltage microprocessor is demonstrated to be feasible. A low swing DC–DC conversion technique is presented that enhances the energy efficiency of a monolithic DC–DC converter. Device reliability issues in monolithic DC–DC converters operating at high input voltages are discussed. Advanced cascode bridge circuits that guarantee the reliable operation of deep submicrometer MOSFETs without exposure to high voltage stress while operating at high input and output voltages are introduced.

An important technique for reducing the impact of supply voltage scaling on circuit performance is scaling the threshold voltages. Exponentially increasing subthreshold leakage currents and worsening short-channel effects at reduced threshold voltages are discussed. Increasing performance degradation caused by die-to-die and within-die parameter variations at reduced gate lengths and threshold voltages is described. Multiple threshold voltage CMOS circuits offer decreased subthreshold leakage currents and enhanced performance by selectively lowering the threshold voltages along speed–critical paths. Dynamic threshold voltage scaling techniques reduce the deleterious effects of standard static threshold voltage scaling. A variable threshold voltage CMOS circuit technique for simultaneously enhancing the speed and power characteristics of dynamic circuits is introduced. Both reverse and forward body bias techniques are applied to domino logic circuits for enhanced robustness against on-chip noise. A circuit technique using sleep switch dual threshold voltage domino logic that provides significant savings in subthreshold leakage energy is described.

Due to lagging battery technologies, increasing cost of cooling, and decreasing yield (caused by degradation in device, circuit, and system-level reliability), the authors of this book strongly believe that the end of the road for traditional speed-centric CMOS design techniques is quickly approaching. Low-power and reliability concerns will dominate at all levels of the design hierarchy and mark the end of this speed-centric road that has been traveled by the semiconductor industry for more than forty years. Meanwhile, market demand for integrated circuits with ever higher performance offering a wider variety of applications will continue to grow consistent with the evolution and increasing complexity of human society. Low-power and reliable integrated circuit and system design will develop into an increasingly exciting field full of opportunities. The concepts presented in this book can be considered as a prelude to a larger discussion of the many possible opportunities for moving the performance and functionality of nanometer semiconductor technologies to even higher levels while staying within a manageable envelope of power consumption and reliability.

Acknowledgments

The authors would like to thank, Dr. Siva G. Narendra of Tyfone, Dr. Vivek K. De, Dr. Tanay Karnik, Dr. Gerhard Schrom, Dr. Peter Hazucha, and Donald Gardner of Intel Corporation, Prof. David Albonesi of Cornell University, Dr. Stephen Dropsho of the Swiss Federal Institute of Technology, and Dr. Radu M. Secareanu of Freescale Semiconductor Corporation for their contributions to this book.

The authors would also like to thank the staff at Wiley who helped make this book happen, specifically, Simone Taylor, Kelly Board, Emily Bone, Lucy Bryan, and Neetu Kalra. It has been great working with each one of you.

Much of the research described in this book was supported in part by the Semiconductor Research Corporation under Contract No. 2003-TJ-1068, the DARPA/ITO under AFRL Contract F29601-00-K-0182, the National Science Foundation under Contract No. CCR-0304574, the Fulbright Program under Grant No. 87481764, grants from the New York State Office of Science, Technology & Academic Research to the Center for Advanced Technology – Electronic Imaging Systems and to the Microelectronics Design Center, and by grants from Xerox Corporation, IBM Corporation, Intel Corporation, Lucent Technologies Corporation, and Eastman Kodak Company.

1 Introduction

The scaling of semiconductor process technologies has been continuing for more than four decades. Advancements in process technologies are the fuel that has been moving the semiconductor industry. In response to growing customer demand for enhanced performance and functionality at reduced cost, a new process technology generation has been introduced by the semiconductor industry every two to three years during the past four decades [1]. Both the performance and the complexity of integrated circuits have grown dramatically since the invention of the integrated circuit in 1959. Microphotographs of the first monolithic integrated circuit (Fairchild Semiconductor, 1959), the first microprocessor (Intel 4004, 1971), and a recent microprocessor (Intel Pentium 4, 2002) are shown in Figure 1.1.

Technology scaling reduces the delay of the circuit elements, enhancing the operating frequency of an integrated circuit (IC) [1]–[5]. The density and number of transistors on an IC are increased by scaling the feature size. By utilizing this growing number of available transistors in each new process technology, novel circuit techniques and microarchitectures can be employed, further enhancing the performance of the ICs beyond the levels made possible by simply scaling (or shrinking) a previous generation [1]–[7]. The price for these performance and functional enhancements has traditionally been increased design complexity and power consumption. The generation, distribution, and dissipation of power are now at the forefront of current problems faced by IC designers.

Historically, circuit techniques and architectures employed during the evolution of the IC have followed two different paths. For a group of technologies, enhancing speed has been at the core of the design process. This class of ICs represents the high end of the performance spectrum. In this high end arena, increasing clock frequency and die size and the widespread use of power-hungry circuit techniques and microarchitectures (with continuously increasing levels of speculative execution often translated into an inefficient use of energy) have increased power consumption many fold over the years [2], [3], [7]. Until recently, the removal of heat in high performance ICs was handled by inexpensive packaging solutions, passive heat sinks, and air fans. With the power dissipation of ICs rising well above 100 W, however, more expensive packaging and cooling solutions such as liquid cooling or refrigeration hardware will soon be required [2]–[10]. Issues related to power dissipation and heat removal are likely to be the primary cause of the end to the trend of continuously decreasing price to performance ratios of high performance ICs.

Multi-Voltage CMOS Circuit Design V. Kursun and E. Friedman
© 2006 John Wiley & Sons, Ltd

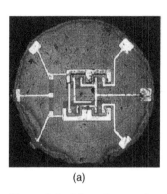
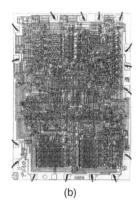
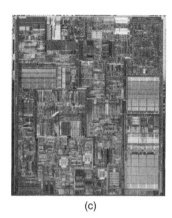

(a)   (b)   (c)

Figure 1.1   Microphotographs of three landmark ICs from the evolution of the IC technology (the sizes of the dies are not to scale). (a) The first monolithic integrated circuit, Fairchild Semiconductor (1959). (b) The first microprocessor, Intel 4004 (1971). (c) A recent Intel Pentium 4 microprocessor (2002)

Another important group of ICs has emerged as a result of customer demand for miniaturization and portability. Portable devices, until recently, represented the low end of the performance spectrum with power constraints always dominating over speed [4], [6], [9]. Extended battery life and reduced system cost constraints drove the portable equipment design process until the 1990s. However, since the 1990s, strong customer demand has been growing for higher performance (for high speed computing and data transfer) and a wider variety of applications in portable equipment. Today, people expect from their portable devices almost the same computing capability as a desktop system.

While the performance of mobile devices continues to advance at a fast pace in accordance with general semiconductor technology trends, the evolution of battery technologies has progressed at a much slower pace [4], [9], [11]. Before rechargeable battery technologies evolved to offer sufficient energy in a miniaturized volume, standard disposable alkaline battery technology was the popular power solution. Frequent battery purchases coupled with the inconvenience of carrying replacement batteries increased the market demand for a rechargeable battery solution. Nickel–cadmium (Ni–Cd) chemistry (invented in 1899 [11]) became the battery supply for portable devices toward the end of the 1980s. Ni–Cd was replaced by nickel–metal–hydride (Ni–M–H) chemistry during the mid-1990s. Ni–M–H batteries offer twice the energy density with faster charging times as compared to Ni–Cd batteries [11]. Lithium-ion (Li-ion) batteries (first introduced in the early 1990s) gradually replaced the Ni–M–H technology toward the end of the last decade. Li-ion, with enhanced energy density characteristics as compared to both Ni–Cd and Ni–M–H batteries, is the most widely used battery technology today [11].

Vendors have responded to the continuous market demand for greater functionality and higher processing speed while continuing to decrease the physical size and weight of portable devices. Batteries are, therefore, required to offer increasing amounts of energy while occupying smaller volumes as semiconductor technologies progress [11]. Today, the lack of a low cost, small volume, and lightweight battery technology with a higher energy density as compared to the Li-ion technology is a primary limitation to further advancements in portable IC technologies.

Traditional circuits and architectures in high performance ICs, because of the power-hungry characteristics of these technologies, are not applicable to those ICs designed for portable systems. Alternatively, circuits and architectures that have been developed for portable devices, because of the typical low throughput characteristics of these technologies, are not effective in high performance ICs. Today, the IC industry is experiencing a shift in requirements at both the high performance and portability ends of the market. Power dissipation is no longer a secondary issue in high performance ICs. Similarly, enhancing throughput is as important as lowering the power, area, and weight in many portable devices. Energy-efficient semiconductor devices, circuit techniques, and microarchitectures are necessary to maintain the pace of expansion that the semiconductor industry has been enjoying for the past forty years [1]–[12].

In retrospect, the invention of the transistor in 1947 can be seen as the first step toward low power electronics. Operation of a vacuum tube requires hundreds of volts of anode voltage and a few watts of power. In comparison, a transistor operates at a higher speed and at a significantly lower supply voltage and consumes orders of magnitude smaller power. Similarly, the invention of the IC in the late 1950s can be seen as the first step toward low power microelectronics. ICs consume less power, are lower weight, and occupy smaller volume while offering the same functionality, with enhanced performance and reliability as compared to circuits composed of discrete devices [13], [14]. These trends that shaped the evolution of IC technology are reviewed in Section 1.1. An outline of this book is summarized in Section 1.2.

1.1 EVOLUTION OF INTEGRATED CIRCUITS

The monolithic IC was invented in 1959. The primary reasons for implementing certain functions as ICs were to lower the weight and size while enhancing the reliability and performance characteristics as compared to circuits composed of discrete devices [13]. ICs were an expensive technology during the 1960s, limiting the use of ICs to specific military applications with severe requirements of weight, size, and reliability. Gordon Moore noticed in 1965, only six years after the birth of the very first IC, that the unit costs of ICs were steadily decreasing as technology evolved and fabrication techniques matured [13], [14]. Moore saw that shrinking transistor sizes, increasing manufacturing yield, and larger wafer and die sizes would make ICs increasingly cheaper, more powerful, and more plentiful. As Moore declared in 1965, 'the future of integrated electronics turned out to be the future of electronics itself' [13]. Advances in IC technology enabled the so-called 'information age' that is experienced today. A timeline of some of the key events that led to the invention and advancements of IC technologies is provided in Figure 1.2.

The general form of Moore's law is depicted in Figure 1.3 [13]. As more components are added to an IC at a particular process technology generation (or technology node), the relative manufacturing cost per component decreases (assuming that the same semiconductor material and the same package are used to incorporate additional components) [13]. However, as more components are integrated onto the same die, the complexity (at the circuits, physical design, and process levels) increases, degrading yield. There is, therefore, an optimum number of components per IC that minimizes the total manufacturing cost at any generation in the evolution of an IC technology [13]. The unit price of a transistor decreases as the device dimensions scale, defect densities are reduced, and wafer and die sizes grow [13], [14]. The

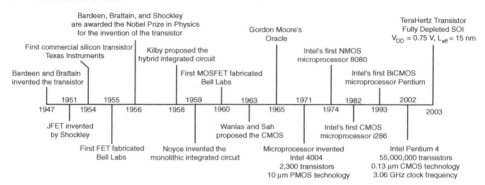

Figure 1.2   A timeline of some of the key events during the evolution of semiconductor technologies

optimum number of transistors that minimizes the total manufacturing cost, therefore, increases from one technology generation to the next as shown in Figure 1.3. The total number of transistors that can be integrated onto a piece of semiconductor material has increased by more than a million times since the mid-1960s, verifying the trends Moore observed in 1965. What began as an observation has become both the compass and engine, setting the bar for the semiconductor industry over the past four decades.

High performance microprocessors currently represent the front end of the market demand for enhanced performance and functionality. No IC technology has witnessed the employment of more aggressive semiconductor process technologies, circuits, and architectures as compared to high performance microprocessors [1], [2]. The high performance micropro-cessor and high density random access memory (RAM) industries have, historically, led the advances in semiconductor technology and hence encountered the side effects of the technology evolution before any other portion of the semiconductor industry.

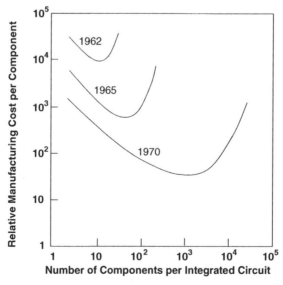

Figure 1.3   The general form of Moore' law [13]

Table 1.1   Technological Trends of High Performance Microprocessors

Vendor	DEC Alpha			AMD			IBM/MOTOROLA PowerPC		
Microprocessor	21064	21164	21264	K5	K6	K7	750	7400	7450
Technology (μm)	0.75	0.5	0.35	0.5	0.3	0.25	0.29	0.2	0.18
Frequency (MHz)	200	300	600	75	233	500	266	400	667
Die area (mm^2)	234	299	314	251	162	184	67	83	106
Transistor count (millions)	1.68	9.3	15.2	4.3	8.8	22	6.35	10.5	33
Supply voltage (V)	3.3	3.3	2.2	3.5	3.3	1.6	2.6	1.8	1.6
Supply current (A)	9.1	15.2	32.7	3.3	9.2	26.3	3	6.3	11.9
Power (W)	30	50	72	11.6	30.2	42	7.9	11.3	19
Power Density (W/cm^2)	12.8	16.7	22.9	4.6	18.6	22.8	11.8	13.6	17.9

The focus of this section is on the advancements of high performance microprocessor technologies. The technological trends in the evolution of the lead Intel microprocessors will be examined. The choice of the lead microprocessor product line of Intel Corporation is due to the significant role that the company has played in the semiconductor industry during the past 35 years. Similar technological trends can also be observed in other leading vendor product lines. Common trends in the characteristics of some of the technological parameters among different microprocessor generations for three vendors are listed in Table 1.1.

The primary force shaping the IC evolution is the advancing fabrication technology that permits technology scaling [1]–[9]. The feature size of the transistors and interconnect have continually been scaled, increasing the integration density in each new process technology generation. The minimum feature size of the transistors in the lead Intel microprocessors has decreased from 10 μm in 1971 to 0.13 μm in 2002 as shown in Figure 1.4. The second primary development behind the IC evolution is the reduction of defect densities due to the maturing fabrication technology, thereby making larger dies (individual ICs or chips) economical. Die areas have grown steadily by about 14% per year from 1971 to 1995 as shown in Figure 1.5. Starting in the mid-1990s, however, limits to further increases in die size became necessary due to concerns about increasing power consumption and soaring fabrication and packaging costs [3], [5]. As a result of the reduced physical dimensions of the transistors and the increased die area, the total number of transistors in the lead Intel microprocessors has increased by twenty four thousand times over the past three decades, as shown in Figure 1.4.

The increasing number of transistors per IC in each new process technology generation offers more tools for enhancing circuit performance and functionality. The propagation delays are reduced as the physical dimensions of the transistors are scaled. Enhancements related to technology scaling coupled with advances in circuits and microarchitectures (such as deeper pipelining, superscalar, and out-of-order execution) have significantly increased the performance of ICs [1]–[8]. As shown in Figure 1.6, the operating frequency of the lead Intel microprocessors has increased by more than twenty eight thousand times since the introduction of the first microprocessor (Intel 4004) in 1971.

Ideal scaling theory suggests shrinking all of the voltages, currents, and physical dimensions and increasing all of the doping concentrations by the same scaling factor (λ) to maintain constant electric fields within a device [40], [56]. Historically, however, the voltages and currents have been scaled at a lower rate as compared to the physical dimensions. The electric

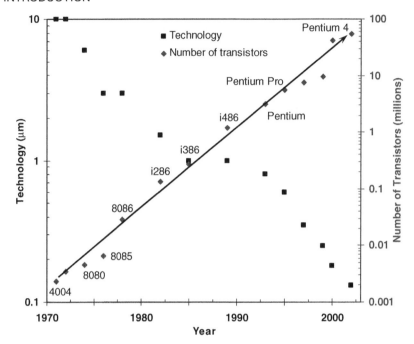

Figure 1.4 Scaling of the minimum feature size and the increasing total number of transistors within each lead Intel microprocessor

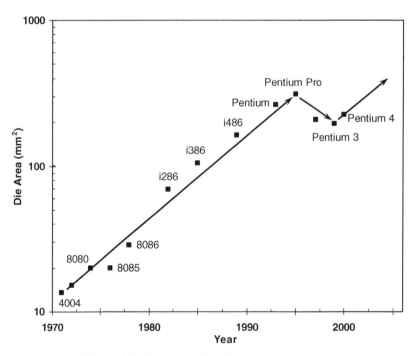

Figure 1.5 Die area of lead Intel microprocessors

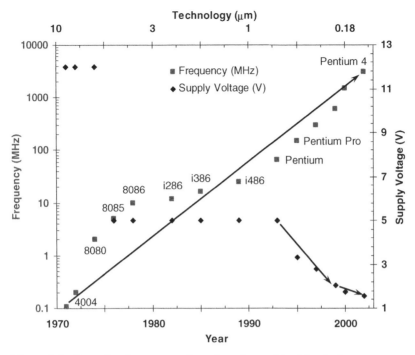

Figure 1.6   Operating frequency and supply voltage of lead Intel microprocessors

fields within the devices have, therefore, significantly increased. An important reason for the reluctance to scale the voltages and currents as rapidly as the physical dimensions has been the beneficial effect of increasing electric fields on device performance [40], [56]. An equally important reason for the slow pace of supply voltage scaling has been the need to maintain high noise margins for maintaining reliability in a difficult-to-control noisy on-chip environment.

The period of technology scaling, since the invention of the first IC, is divided into two primary eras depending upon the characteristics of the supply voltage in a scaled technology as compared to a preceding technology generation. The supply voltage in the first three Intel microprocessor generations was 12 V as shown in Figure 1.6. Starting with the 3 μm technology node, the supply voltage was reduced to 5 V. IC supply voltages were maintained at 5 V until the 0.8 μm technology node was commercialized during the early 1990s (see Figure 1.6). At the 0.8 μm technology node, supply voltage scaling became an essential part of the technology scaling process due to transistor reliability and power consumption concerns [3]–[5], [15]. The era (until 1993 in the case of Intel) during which supply voltage scaling was not necessarily a part of technology scaling is called the constant voltage scaling era. The technology scaling era (after 1993 in the case of Intel), during which supply voltage scaling occurs with scaling of the other device parameters, is called the constant field scaling era [3], [4], [15]. Constant field scaling arises from the concept that the supply voltage for a new technology is ideally chosen to maintain constant electric fields between the terminals of a transistor [15]. The need to slow the rate of increase in power consumption became an increasingly important factor in supply voltage scaling toward the end of the 1990s. Today, the requirements for lowering the power consumption and improving device reliability

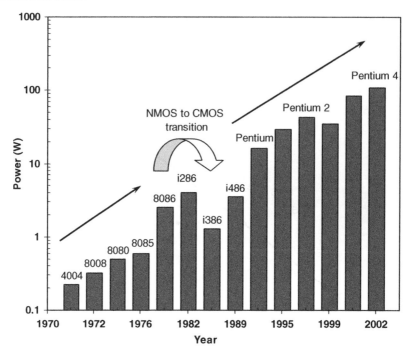

Figure 1.7   Maximum power consumption of lead Intel microprocessors

together with circuit speed determine the rate of supply voltage scaling in each new technology generation [1]–[5], [7], [8], [11], [15], [16].

An increase in the operating frequency and die size (due to the greater number of transistors for the additional circuitry and novel microarchitectures) not only enhances the speed, but also increases the power consumption [1]–[6], [8]–[10], [16], [17]. As shown in Figure 1.7, the power consumption of the lead Intel microprocessors has been increasing over the past 30 years. The technology in the first two Intel microprocessor generations was p-channel metal oxide semiconductor (PMOS). Starting with the Intel 8080, n-channel metal oxide semiconductor (NMOS) became the preferred technology due to the speed and area advantages of NMOS transistors as compared to PMOS transistors. NMOS circuits, however, suffered from higher static DC power consumption and lower noise margins [18], [20]. By the end of the 1970s, scaling of NMOS technology became increasingly difficult as the low noise margins of the NMOS circuits did not permit supply voltage scaling to accompany scaling of the feature size [18]. The increasing number of transistors operating at higher clock frequencies at a high supply voltage coupled with the intrinsic static DC power consumption of the NMOS circuits set the stage for the end of a decade-long dominance of NMOS as the technology of choice. As shown in Figure 1.8, the power density of the last NMOS Intel microprocessor (the i8086 that was commercialized in 1978) is similar to the power density of a kitchen hot plate. The packaging and cooling technologies available at the beginning of the 1980s were quite limited, permitting no further technological advances that would lead to an increase in power dissipation.

The complementary metal oxide semiconductor (CMOS) circuit topology (first proposed in 1963 by Wanlass and Sah [165]) was adapted by the IC industry in the early 1980s due to the

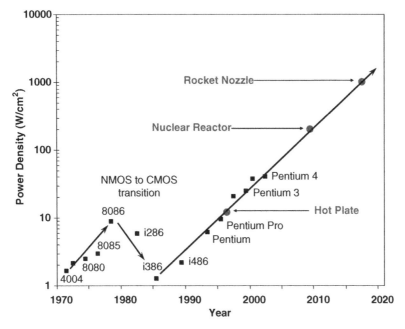

Figure 1.8 Power density trends of lead Intel microprocessors

intrinsically lower power consumption and enhanced scaling characteristics of CMOS as compared to NMOS [18]–[20]. The higher noise margins in CMOS circuits made possible supply voltage scaling that accelerated in the 1990s, enhancing both transistor reliability and energy efficiency. CMOS became the preferred circuit topology in the lead Intel micro-processors starting with the i286 (introduced in 1982). The transition from NMOS to CMOS reduced both the power consumption and the power density of the Intel microprocessors as shown in Figures 1.7 and 1.8, respectively [5].

The reduction in the power dissipation of high performance microprocessors due to the transition to CMOS, however, provided only temporary relief. Maintaining the approach of employing higher clock frequencies coupled with power-hungry circuits and highly spec-ulative architectures in order to achieve enhanced performance, the power consumption and power density of the post-NMOS era (i.e., CMOS and BiCMOS) ICs were, once again, pushed to higher levels. As illustrated in Figures 1.7 and 1.8, respectively, both the power consumption and power density of the lead Intel microprocessors (with the exception of the first generation Pentium 3) have been increasing since the introduction of the second generation CMOS microprocessor (i386) in 1985. As depicted in Figure 1.8, the power density of current high performance microprocessors has greatly exceeded the power density of the heating coil of a kitchen hot plate [2], [3], [5], [21].

The temperature of a die is controlled to maintain proper operation of the circuitry compliant with technical specifications [5], [10], [22]. Thermal management of high performance ICs has become increasingly difficult due to the continuously increasing power dissipation and power density in each new process technology generation [2]–[5], [10], [12], [16], [17], [21], [23]. Within a few technology generations, traditional cooling solutions such as low cost heat sinks and air flow fans will become ineffective for thermal management [2]–[5], [10], [22]. If the current trend in the rate of increase in the power levels continues, ICs

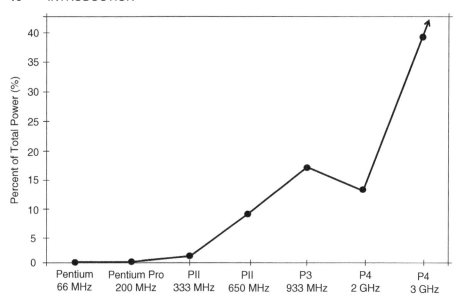

Figure 1.9 Increasing contribution of leakage currents to the total power consumption of the lead Intel microprocessors [161]

will consume thousands of watts of power in the near future [2], [5], [21]. The power density of a high performance microprocessor will, within the next decade, exceed the power density levels encountered in typical rocket nozzles [2], [5]. Low cost cooling solutions that can handle power densities in excess of nuclear reactors or rocket nozzles do not presently exist for ICs. As acknowledged by many designers and researchers, excessive power dissipation has emerged as the single greatest jeopardy to further advances in IC technologies [1]–[14].

Dynamic switching power consumption has typically been the dominant source of power consumption in CMOS ICs. Recently, however, leakage power has become a significant portion of the total power consumption in high complexity CMOS ICs, as illustrated in Figure 1.9. Ideally, an MOS switch has infinite input impedance. Similarly, an ideal cut-off transistor has infinite drain-to-source resistance. However, in reality, an active transistor has a finite input impedance and a cut-off transistor has a finite channel resistance, producing gate oxide and subthreshold leakage current, respectively. Due to the aggressive scaling of the threshold voltages and the thickness of the gate dielectric layer in order to enhance device speed, modern MOSFETs no longer resemble, even remotely, an ideal switch. As illustrated in Figure 1.9, subthreshold and gate oxide leakage currents will become the dominant source of power consumption in the near future.

Another important challenge directly linked to the advances of semiconductor technologies is maintaining the reliability of scaled CMOS circuits. The reliability of CMOS ICs has degraded due to scaling the device and interconnect dimensions and the on-chip voltage levels. Error-free operation of CMOS circuits has become increasingly challenging as IC technologies evolve. CMOS ICs have become more sensitive to noise while on-chip noise levels continue to rise with each new technology generation. Various sources of noise in a microprocessor are schematically illustrated in Figure 1.10. The clock distribution network acts as a source of noise to the surrounding circuitry and interconnect lines. On-chip clock generators inject considerable amounts of noise into the substrate. Similarly, a monolithic

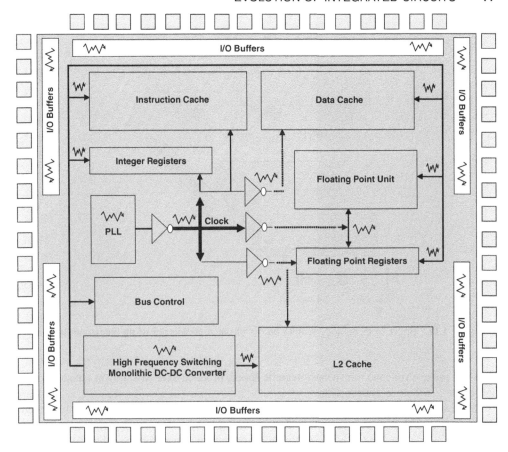

Figure 1.10   Various sources of noise in a microprocessor

switching DC–DC converter can produce significant noise on a microprocessor die, as illustrated in Figure 1.10.

An important source of noise in CMOS ICs is interconnect coupling noise [158], [159]. Due to increasing device densities, interconnect lines are physically closer with each new technology generation, as illustrated in Figure 1.11. The resistance of the interconnect lines increases as the width of the interconnect lines is reduced with technology scaling. Due to the increasing resistance of the interconnect lines, the delay due to the interconnect rather than the gate delays dominates the propagation delay characteristics in current CMOS circuits, as illustrated in Figure 1.12. The higher interconnect resistance also increases the parasitic power dissipation. In order to limit the increased resistance of the interconnect lines, the height of the interconnect lines is scaled at a much smaller rate as compared to the width with each new technology generation. The aspect ratio, therefore, increases significantly, thereby increasing the coupling capacitance between adjacent interconnect lines on the same metal layer. Similarly, due to the vertical scaling trend of adding more metal layers to CMOS fabrication processes, the intermetal layer coupling capacitances also tend to increase. Noise generated on a quiescent line (victim line) due to the coupling capacitance with a nearby line (aggressor

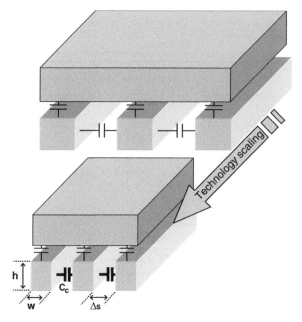

Figure 1.11   Effect of technology scaling on the physical geometries of the interconnect lines

line) can cause erroneous transitions, degrade speed, produce excessive power consumption, and cause a circuit to malfunction.

The power distribution network is another important source of noise in deeply scaled nanometer CMOS ICs. An important factor exacerbating the on-chip noise issue is the increasing current demand of modern ICs. While the power consumption of the ICs continues to increase, the supply voltages have been reduced, as shown in Figure 1.6. The supply current, therefore, increases, as shown in Figure 1.13. Increased current demand of ICs

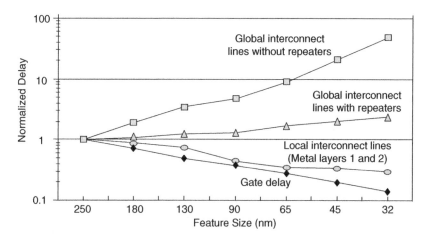

Figure 1.12   Effect of technology scaling on interconnect and gate delays [162]

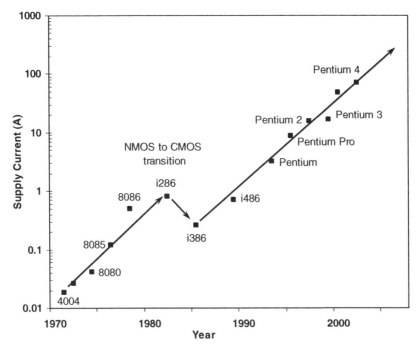

Figure 1.13   Increasing current demand of lead Intel microprocessors

coupled with scaled wire dimensions create metal migration and voltage drop problems within the power distribution network [21], [24], [25].

Power supply noise has both low frequency and high frequency components [158]. The low frequency component of the power supply noise is due to the resistive IR drops on the printed circuit board, within the package, and along the on-chip power grid. The tolerance of ICs to voltage fluctuations in the power supply grid is typically reduced while the resistance of the interconnect is increased with technology scaling [18], [19], [24]. The resistive voltage drop of the power distribution grid, therefore, has become an increasingly important concern for maintaining performance and reliability. Alternatively, the high frequency components of the power supply noise are due to the inductance of the printed circuit board planes, package, and the on-chip power grid. Current slew rates typically increase due to the higher operating frequencies as well as the growing current demand in each new technology generation. Simultaneous switching noise $(L\,di/dt)$ due to the inductance of a power distribution grid affects the supply voltage, thereby degrading the performance and possibly causing circuit malfunctions [25], [164].

Power generation, delivery, and dissipation are primary limitations to further advancements of IC technologies [1]–[9], [16]–[18], [21], [23]. In order to continue to reduce the unit cost of an IC while simultaneously enhancing the performance and functionality, radical changes are required in the way ICs have been designed during the past three decades. Higher performance at all costs is no longer an option. Novel energy-efficient devices, circuits, microarchitectures, and macroarchitectures must be developed to lower the rate of increase in the power consumed by next generation ICs.

1.2 OUTLINE OF THE BOOK

Several new techniques for the design of low power and high performance ICs are described in this book. Particular emphasis is placed on issues related to the scaling of the supply and threshold voltages in high performance ICs.

An analysis of power dissipation-related problems faced by the semiconductor industry starts with identifying the sources of power consumption. The primary sources of power consumption in CMOS ICs are described in Chapter 2. Specifically, dynamic, short-circuit, leakage, and static DC power components are individually described.

Supply and threshold voltage scaling techniques, aimed at lowering power consumption and enhancing device reliability without degrading performance, are discussed in Chapter 3. The importance of supply voltage scaling is discussed from an energy efficiency point of view. As the supply voltage is reduced, the performance of an IC degrades due to reduced transistor currents [27]. Systems with multiple supply voltages can minimize the degradation in speed while reducing power by selectively lowering the supply voltages along non-critical delay paths [28]. Dynamic and static versions of multiple supply voltage IC design techniques are reviewed. Another alternative technique for reducing the impact of supply voltage scaling on circuit performance is threshold voltage scaling. During the past decade, threshold voltage scaling has accelerated together with scaling of the supply voltages. At reduced threshold voltages, however, subthreshold leakage currents increase. Supply voltage scaling when coupled with threshold voltage reduction, therefore, increases the leakage power while reducing the dynamic switching power. Multiple threshold voltage circuits reduce leakage currents while enhancing performance by selectively lowering the threshold voltages only on speed-critical paths [29]. Dynamic threshold voltage scaling (V_t-hopping) and multiple threshold voltage CMOS circuit techniques are reviewed in Chapter 3. Dynamic and static versions of multiple threshold voltage circuit techniques are also discussed in this chapter.

A significant issue with threshold voltage and device scaling is the increasing effect of die-to-die and within-die parameter variations on the speed and power dissipation characteristics of CMOS ICs. Die-to-die and within-die fluctuations of the critical dimensions (gate length, gate oxide thickness, and junction depletion width) effectively increase with technology scaling. Moreover, the sensitivity of the threshold voltage to variations in the critical dimensions is greater due to increasing short-channel effects as the gate length and threshold voltage are both reduced with technology scaling. Process variations cause ICs to exhibit different speed and power characteristics. The electrical characteristics of a CMOS circuit fabricated in a deep submicrometer process technology have become increasingly probabilistic (less deterministic). The number of individual dies that satisfy a target clock frequency and maximum power dissipation constraint is lower, degrading the yield. The increasing cost of fabricating deep submicrometer ICs is, therefore, further aggravated by lower yields caused by greater process variations. Challenges imposed by these parameter variations are also addressed in Chapter 3.

The generation and distribution of the energy required for the proper functioning of an IC are important challenges due to system-level power budget limitations and circuit reliability issues. Increasing supply currents together with reduced supply voltages degrade the energy efficiency and voltage quality of power generation and distribution networks in high performance ICs. Energy-efficient low voltage monolithic DC–DC conversion and voltage regulation techniques are developed in this book. Before presenting the details of these monolithic DC–DC conversion techniques in the following chapters, a basic background to DC–DC conversion and a review of several widely employed types of low voltage DC–DC converters are presented in Chapter 4.

In single power supply microprocessors, the primary power supply is typically an external (non-integrated) buck converter. In a typical non-integrated switching DC–DC converter, significant energy is dissipated by the parasitic impedances of the interconnect among the non-integrated devices (the filter inductor, filter capacitor, power transistors, and pulse width modulation circuitry) [9], [26], [30], [31]. Moreover, the devices of a discrete DC–DC converter are typically fabricated in older technologies with poor parasitic impedance characteristics. Integrating a DC–DC converter onto a microprocessor can potentially lower the parasitic losses as the interconnect between (and within) the DC–DC converter and the microprocessor is reduced. Additional energy savings can be realized by utilizing advanced deep submicrometer fabrication technologies with lower parasitic impedances. The efficiency attainable with a monolithic DC–DC converter can therefore be higher than a non-integrated DC–DC converter [30]. An analysis of on-chip buck converters is presented in Chapter 5. A model of the parasitic impedances of a buck converter is described. With this model, a design space is determined that supports the integration of active and passive devices on the same die for a target technology. A monolithic, high efficiency, and high frequency switching DC–DC converter with an integrated inductor on the same die as a dual supply voltage microprocessor is demonstrated to be feasible.

The model presented in Chapter 5 provides an accurate representation of the parasitic losses of a full voltage swing DC–DC converter (with an error of less than 2.4% as compared to simulation). A high switching frequency is the key design parameter that enables the full integration of a high efficiency DC–DC converter. At these high switching frequencies, the energy dissipated in the power MOSFETs and gate drivers dominates the total losses of a DC–DC converter. The efficiency can, therefore, be enhanced by applying a variety of MOSFET power reduction techniques [31]. A low swing MOSFET gate drive technique is described in Chapter 6 that enhances the efficiency of a DC–DC converter. An advanced circuit model for low swing circuit optimization is also presented. The gate voltages and transistor sizes are included as independent parameters in this model. The optimum gate voltage swing of a power MOSFET that maximizes efficiency is shown to be lower than a standard full voltage swing. Lowering the input and output voltage swing of a power MOSFET gate driver is shown to be effective for enhancing the efficiency characteristics of a DC–DC converter.

Due to the advantages of high voltage power delivery on a circuit board and monolithic DC–DC conversion, next generation low voltage and high power microprocessors are likely to require high input voltage, large step-down DC–DC converters monolithicly integrated onto the same die. The voltage conversion ratios attainable with standard non-isolated switching DC–DC converter circuits are limited, however, due to MOSFET reliability issues. Provided that a DC–DC converter is integrated onto the same die as a microprocessor (fabricated in a low voltage nanometer CMOS technology), the range of input voltages that can be applied to a standard DC–DC converter circuit is further reduced. A standard non-isolated switching DC–DC converter topology such as the buck converter circuit discussed in Chapters 5 and 6 is, therefore, not suitable for providing a high voltage conversion ratio in future high performance ICs. Three cascode bridge circuits that can be used in monolithic DC–DC converters that provide high voltage conversion ratios are presented in Chapter 7. The circuits ensure that the voltages across the terminals of all of the MOSFETs in a DC–DC converter are maintained within the limits imposed by available low voltage CMOS technologies.

In ICs with multiple supply voltages, signal transfer among the regions operating at different voltage levels requires specialized voltage interface circuits [32]. Another low power circuit technique that requires voltage-level conversion is low swing interconnect signaling. At each new IC generation, the relative amount of interconnect increases due to the greater number of transistors and the larger die size. In many recent systems, charging and discharging these interconnect lines can require more than 50% of the total power consumed on-chip. In certain programmable logic devices, more than 90% of the total power consumption is due to the interconnect wires [32]. Decreasing the signal voltage swing on the interconnect can significantly decrease the power consumption. In a low swing interconnect architecture, voltage-level converters are placed at the driver and receiver ends of the low swing interconnect to change the voltage levels. A bidirectional CMOS voltage interface circuit that drives high capacitive loads to full swing at high speed while consuming no static DC power is presented in Chapter 8. The propagation delay, power consumption, and power efficiency characteristics of this voltage interface circuit are compared to other interface circuits described in the literature. The voltage interface circuit offers significant power savings and lower propagation delay.

Domino logic circuit techniques are extensively applied in high performance micropro-cessors due to the superior speed and area characteristics of dynamic CMOS circuits as compared to static CMOS circuits. High speed operation of domino logic circuits is primarily due to the lower switching threshold voltage of dynamic circuits as compared to static gates. This property of a lower switching threshold voltage, however, makes domino logic circuits highly sensitive to noise as compared to static gates. On-chip noise becomes more severe with technology scaling and higher operating frequencies. Furthermore, the noise sensitivity of domino logic circuits increases with technology scaling. Error-free operation of domino logic circuits has, therefore, become a major challenge [33]. A variable threshold voltage keeper circuit technique is presented in Chapter 9 for simultaneous power reduction and speed enhancement of domino logic circuits. The threshold voltage of the keeper transistor is dynamically modified during circuit operation to reduce the contention current without sacrificing noise immunity. The variable threshold voltage keeper circuit technique is shown to enhance circuit evaluation speed by up to 60% while reducing power consumption by 35% as compared to a standard domino logic circuit. The keeper size can be increased while preserving the same delay or power characteristics as compared to a standard domino circuit since the contention current is reduced with this technique. The domino logic circuit technique offers 14.1%, 8.9%, or 11.9% higher noise immunity under the same delay, power, or power–delay product conditions, respectively, as compared to the standard domino logic circuit technique. Forward body biasing the keeper transistor is also described for improved noise immunity as compared to a standard domino circuit with the same keeper size. It is shown that by applying forward and reverse body bias circuit techniques, the noise immunity and evaluation speed of domino logic circuits are both enhanced.

The subthreshold leakage current of a domino logic circuit can vary dramatically with the voltage state of the dynamic and output nodes. A quantitative review of the subthreshold leakage current characteristics of standard low threshold voltage and dual threshold voltage domino logic circuits is presented in Chapter 10. Different subthreshold leakage current conduction paths which exist depending upon whether the dynamic node is charged or discharged are identified. A discharged dynamic node is preferable for reducing leakage current in a dual threshold voltage circuit. Alternatively, a charged dynamic node is preferred for lower subthreshold leakage current in a standard low threshold voltage domino logic circuit with stacked pull-down devices, such as an AND gate. The effect of dual threshold voltage CMOS technologies on the noise immunity characteristics of domino logic circuits is also evaluated in Chapter 10.

A circuit technique is presented in Chapter 11 for exploiting the dynamic node voltage-dependent asymmetry of the subthreshold leakage current characteristics of domino logic circuits. Sleep switch transistors are used to place an idle dual threshold voltage domino logic circuit into a low subthreshold leakage state. The circuit technique enhances the effectiveness of a dual threshold voltage CMOS technology to reduce subthreshold leakage current by strongly turning off all of the high threshold voltage transistors. The sleep switch circuit technique significantly reduces the subthreshold leakage energy as compared to both standard low threshold voltage and dual threshold voltage domino logic circuits. A domino adder enters and leaves the low leakage sleep mode within a single clock cycle. The energy overhead of the circuit technique is low, justifying the activation of the sleep scheme by providing a net saving in total power consumption during short idle periods.

A summary of the themes and ideas presented in this book is provided in Chapter 12. It is emphasized that low power and reliability concerns will dominate at all levels of the design hierarchy as the end of the traditional speed-centric methodology approaches. Some of the opportunities that exist for low power and reliable IC and system design are revisited.

2 Sources of Power Consumption in CMOS ICs

Power consumption is a primary limitation to the further advancement of semiconductor technologies. Identifying the sources of power consumption is critical for developing power reduction techniques at the fabrication technology, circuit, and architecture levels. There are four sources of power consumption in CMOS circuits. The total power consumption of a CMOS circuit is

$$P_{total} = P_{dynamic} + P_{leakage} + P_{short\text{-}circuit} + P_{DC}, \tag{2.1}$$

where $P_{dynamic}$ is the dynamic switching power dissipated while charging or discharging the parasitic capacitances during a node voltage transition. $P_{leakage}$ is a combination of the subthreshold leakage power due to the non-ideal off-state characteristics of the MOSFET switches and the gate leakage power caused by carrier tunneling through the thin gate oxides. $P_{short\text{-}circuit}$ is the transitory power dissipated during an input signal transition when both the pull-up and pull-down networks of a CMOS gate are simultaneously on. P_{DC} is the static DC power consumed when a CMOS circuit is driven by low voltage swing input signals.

Each of these four sources of power consumption in a CMOS IC are analyzed in this chapter. The dynamic switching power is discussed in Section 2.1. The sources of leakage power are identified in Section 2.2. The mechanisms of short-circuit and static DC power consumption are discussed in Sections 2.3 and 2.4, respectively.

2.1 DYNAMIC SWITCHING POWER

The dominant component of power consumption in a typical CMOS circuit is the dynamic switching power [4], [9], [12], [21], [23], [27]–[29], [36]. The dynamic switching power is dissipated while charging or discharging the parasitic capacitances during the voltage transitions of the nodes within a CMOS circuit. The dynamic switching power is independent of the type of switching gate and the shape of the input waveform (input rise and fall times). The dynamic switching power is dependent only on the supply voltage, the switching frequency, the initial and final voltages, and the equivalent capacitance of a switching node

Multi-Voltage CMOS Circuit Design V. Kursun and E. Friedman
© 2006 John Wiley & Sons, Ltd

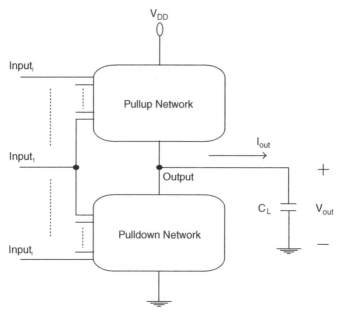

Figure 2.1   A CMOS gate driving an output capacitor. The drain-to-body junction capacitances of the driver gate, the equivalent capacitance of the interconnect, and the gate oxide capacitance of the transistors in the fan-out gates are lumped into a single equivalent load capacitance, C_L

[4], [9], [36]. Since the switching power is independent of the type of switching gate, a block diagram representation of a generic CMOS gate (as shown in Figure 2.1) is used in this section to explain dynamic switching power dissipation in CMOS circuits.

For a low-to-high transition at the output node, the pull-up network is activated and the pull-down network is disabled. The portion of the current sourced by the power supply that passes through the pull-up transistors to charge the output capacitor is denoted by $I_{out}(t)$. The instantaneous power drawn from the power supply to charge the output capacitor is

$$P(t) = V_{DD}I_{out}(t), \tag{2.2}$$

$$I_{out}(t) = C_L \frac{dV_{out}(t)}{dt}, \tag{2.3}$$

where V_{DD} is the power supply voltage and $V_{out}(t)$ is the instantaneous voltage across the output capacitor.

The energy drawn from the power supply for a $V_1 \rightarrow V_2$ transition at the output node voltage is

$$E_{V_1 \rightarrow V_2} = \int_{t_1}^{t_2} P(t)dt = V_{DD}\int_{t_1}^{t_2} I_{out}(t)dt = C_L V_{DD}\int_{V_1}^{V_2} dV_{out}(t) = C_L V_{DD}(V_2 - V_1), \tag{2.4}$$

$$V_{swing} = V_2 - V_1, \tag{2.5}$$

$$E_{V_1 \rightarrow V_2} = C_L V_{DD} V_{swing}, \tag{2.6}$$

where $E_{V_1 \rightarrow V_2}$ is the energy drawn from the power supply to charge the output capacitance from an initial voltage of V_1 to a final voltage of V_2 and t_1 and t_2 are the times for the output voltage to reach V_1 and V_2, respectively. After the $V_1 \rightarrow V_2$ transition of the output node voltage is completed, the energy stored in the output capacitor is

$$E_{C_L} = \int_{t_1}^{t_2} P_{C_L}(t)dt = \int_{t_1}^{t_2} V_{out}(t)I_{out}(t)dt = C_L \int_{V_1}^{V_2} V_{out}(t)dV_{out}(t) = \frac{1}{2}C_L(V_2^2 - V_1^2), \tag{2.7}$$

where $P_{C_L}(t)$ is the instantaneous power stored in the output capacitor. The remaining portion of the energy drawn from the power supply is dissipated in the parasitic resistances of the pull-up network transistors during the output $V_1 \rightarrow V_2$ transition.

For a high-to-low transition of the output node voltage, the pull-up network transistors are cut off and the pull-down network is enabled. The magnitude of the portion of the instantaneous current through the pull-down network transistors that discharges the output node capacitor is $I_{out}(t)$. The polarity (or direction) of this discharging current is opposite to the direction of the load current as shown in Figure 2.1. The energy dissipated in the parasitic resistances of the pull-down network transistors to discharge the output capacitor is

$$E_{V_2 \rightarrow V_1} = \int_{t_1}^{t_2} P_{pulldown}(t)dt = -\int_{t_1}^{t_2} V_{out}(t)I_{out}(t)dt = -C_L \int_{V_2}^{V_1} V_{out}(t)dV_{out}(t), \tag{2.8}$$

$$E_{V_2 \rightarrow V_1} = -\frac{1}{2}C_L(V_1^2 - V_2^2) = \frac{1}{2}C_L(V_2^2 - V_1^2) = E_{C_L}, \tag{2.9}$$

where $E_{V_2 \rightarrow V_1}$ is the energy dissipated in the pull-down network while discharging the output capacitor from an initial voltage of V_2 to a final voltage of V_1 and t_1 and t_2 are the times for the output voltage to reach V_2 and V_1, respectively. As given by (2.7) and (2.9), all of the energy stored in the output capacitor during a $V_1 \rightarrow V_2$ transition is dissipated in the resistances of the pull-down network transistors during the following $V_2 \rightarrow V_1$ transition.

The power is the energy stored or dissipated per unit of time [38]. Assuming that a node voltage periodically transitions between V_1 and V_2 with a period of T_s, the average dynamic power consumed by a CMOS gate driving the switching node is

$$P = \frac{E_{V_1 \rightarrow V_2}}{T_s} = f_s C_L V_{DD} V_{swing}. \tag{2.10}$$

In a CMOS IC, all of the internal nodes do not necessarily change state at each clock cycle. In a synchronous CMOS IC, if statistical data are available for the average number of transitions experienced by a node during the execution of a certain task, an average activity factor α can be introduced into the power and energy expressions. The average power consumed for switching a node i in a CMOS circuit is

$$P_i = \alpha_i f_s C_L V_{DD} V_{swing}, \tag{2.11}$$

where P_i is the average dynamic switching power dissipation of the gate driving the ith node and α_i is the probability that a state changing voltage transition will occur at the ith node within a clock cycle.

Summing the average dynamic switching power consumed for switching all of the nodes within a circuit, the total dynamic switching power consumption of an IC is [9], [27]

$$P_{Total} = f_s V_{DD} \sum_{i=1}^{N} \alpha_i C_{L_i} V_{swing_i}, \qquad (2.12)$$

where N is the total number of nodes within a CMOS circuit, C_{L_i} is the equivalent parasitic capacitance of the ith node, and V_{swing_i} is the voltage swing on the ith node.

In CMOS circuits, the node voltages are typically full swing between ground and V_{DD}. The average switching power consumed by a full swing CMOS gate is (from (2.11))

$$P_i = \alpha_i f_s C_L V_{DD}^2. \qquad (2.13)$$

2.2 LEAKAGE POWER

A transistor switch is essentially a resistive–capacitive network between the power supply and ground. Due to the non-ideal off-state characteristics (a finite resistance) of a transistor, current is drawn from a power supply even when a transistor operates in the cut-off region. The leakage currents are dominated by weak inversion and reverse biased p–n junction diode currents in long channel devices [4], [9], [36], [37], [48]. In deep submicrometer ICs, other leakage mechanisms such as drain-induced barrier lowering (DIBL) and gate oxide tunneling are also important [4], [37], [39]–[48].

The primary mechanisms of leakage current within deep submicrometer ICs are reviewed in this section. The sources of subthreshold leakage current are discussed in Section 2.2.1. The gate insulator leakage current mechanisms in deeply scaled nanometer MOSFETs are reviewed in Section 2.2.2.

2.2.1 Subthreshold Leakage Current

A MOSFET operates in the weak inversion (subthreshold) region when the magnitude of the gate-to-source voltage is less than the magnitude of the threshold voltage [44]. In the weak inversion mode, current conduction between the source and drain (the subthreshold leakage current) is primarily due to diffusion of the carriers [4], [9], [23], [44], [48]. The transistor off-state current (I_{OFF}) is the drain current when the gate-to-source voltage is zero [4], [37]. I_{OFF} is affected by the threshold voltage, channel length, channel width, depletion width beneath the channel area, channel/surface doping profiles, drain/source junction depths, gate oxide thickness, supply voltage, and the junction temperature [37]. The variation of the drain current of an NMOS transistor as a function of the gate voltage for three different drain voltages in a $0.18\,\mu m$ CMOS technology is shown in Figure 2.2. Measurements reveal the dominant leakage current mechanisms within a MOSFET fabricated in a deep submicrometer CMOS process.

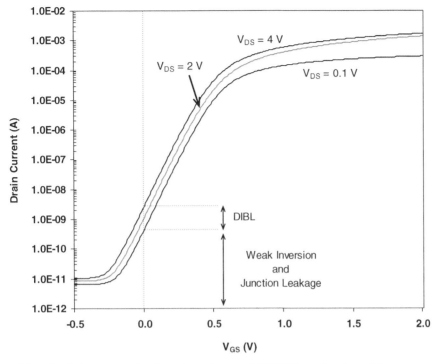

Figure 2.2   The drain current of a short-channel n-type MOSFET as a function of the gate-to-source voltage (V_{GS}) for three different drain-to-source voltages (V_{DS}). The DIBL, weak inversion, and p–n junction diode leakage components of the drain subthreshold leakage current are indicated (0.18 μm CMOS technology, $W = 10W_{min}$, and $L = L_{min}$)

As the length of the channel is scaled, the capability of the gate to control the charge and potential distribution in the channel area degrades. The threshold voltage of a MOSFET is reduced with decreasing channel length [37], [39]–[45], [48]. The effects of scaling the channel length on the threshold voltage and subthreshold leakage current characteristics of a MOSFET are called *short-channel effects*. The short-channel effects are discussed in Section 2.2.1. While the gate loses some control of the channel region, the effect of the drain on the voltage potential distribution across the channel area increases with scaling of the gate length [45], [50]. The effect of the bias conditions of the drain on the threshold voltage and subthreshold leakage current characteristics of a MOSFET is called drain-induced barrier-lowering (DIBL) [44], [48]. DIBL is discussed in Section 2.2.1.2. The various parameters that characterize subthreshold leakage current in a deep submicrometer IC are reviewed in Section 2.2.1.3.

2.2.1.1 Short-Channel Effects

In a long-channel MOSFET, extensions of the space charge regions at the source and drain-to-body p–n junctions into the channel area occupy only a small fraction of the entire channel region. The gate voltage controls most of the space charge induced in the channel area of a long-channel device during inversion. The extensions of the source and drain depletion

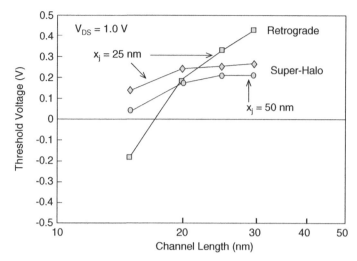

Figure 2.3 Short-channel MOSFET threshold voltage roll-off for super-halo (both vertically and laterally non-uniform) and retrograde (vertically non-uniform) doping profiles [37], [40]–[43]

regions into the channel area, therefore, have a negligible effect on the threshold voltage of a long-channel device [4], [48], [50].

As the channel length is decreased, however, the gate begins to play a small part in the creation of a depletion layer in the channel area [37], [45], [50]. As the channel length of a MOSFET is reduced with technology scaling, the depletion regions around the source and drain terminals become closer. The total depth of the source and drain depletion regions becomes comparable to the effective channel length in deep submicrometer devices. More charge is contributed to the depletion region beneath the gate area by the source-to-substrate and drain-to-substrate depletion layers in a short-channel device as compared to a long-channel device [44], [48], [50]. The threshold voltage of a transistor is, therefore, reduced (typically called V_t roll-off) with decreasing gate length as shown in Figure 2.3 [39]–[45].

The increasing dependence of the threshold voltage of a short-channel MOSFET on the gate length is an issue that threatens the future of technology scaling [40], [50]. The within-die and inter-die variation of circuit parameters such as the gate length causes CMOS circuits to display different clock frequency and leakage power characteristics from die to die, wafer to wafer, and lot to lot. Increasing short-channel effects combined with increasing die-to-die and within-die variations of the critical dimensions (channel length, oxide thickness, and junction depth) cause the performance of CMOS ICs to become increasingly probabilistic, degrading yield. The number of dies that satisfy functional, timing, and power dissipation specifications is reduced with technology scaling, thereby further increasing the cost of fabrication in each new technology generation [50].

Novel device structures are, therefore, necessary to control short-channel effects in deep submicrometer CMOS technologies. As shown in Figure 2.3, a retrograde doping profile (vertical non-uniform doping) in the channel region causes an unacceptably large threshold voltage degradation as the channel length is reduced. Novel and more complex doping profiles such as super-halo (both vertical and lateral non-uniform doping) are needed to control short-channel effects in deep submicrometer devices [40]–[43], [51]–[55].

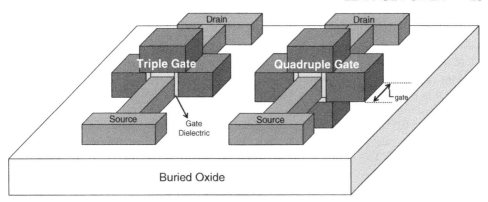

Figure 2.4 Triple gate and gate-all-around (quadruple gate) MOSFETs

The scaling of planar silicon devices has been continuing for approximately five decades. However, due to degradation in device electrical characteristics caused by short-channel effects and fabrication-related difficulties, scaling of standard planar MOSFETs is expected to slow down within the next decade ($L_{\min} \approx 10$ nm). Today, IC technologies are shifting from bulk silicon to silicon-on-insulator (SOI) as the new industry standard [149]–[152]. SOI devices have enhanced short-channel characteristics as compared to standard bulk silicon devices. Moreover, SOI devices typically operate at considerably higher speeds due to lower junction capacitances as compared to bulk silicon devices [149], [150]. Multiple gate variations of SOI-based technologies, as shown in Figure 2.4, are likely to be widely used within the next decade [147], [152]. A gradual shift toward devices based on novel materials, such as carbon nanotubes, is likely to be observed toward the end of the next decade (approximately 2020) [153], [154].

2.2.1.2 Drain-Induced Barrier-Lowering

As the magnitude of the reverse bias voltage across the drain-to-body p–n junction is increased, the depth of the junction depletion layer increases. A deeper depletion layer around the drain contributes a larger amount of depletion charge to the channel. An increased drain-to-body reverse bias voltage, therefore, enhances the short-channel effects and lowers the magnitude of the threshold voltage of a MOSFET. The threshold voltage degradation caused by an increased or decreased drain bias voltage of an n-type or p-type MOSFET, respectively, is commonly referred to as DIBL [37], [44], [48]. As shown in Figure 2.2, a significant portion of the subthreshold leakage current of a deep submicrometer MOSFET can be due to DIBL at high reverse bias voltages across the drain-to-body p–n junction.

2.2.1.3 Characterization of Subthreshold Leakage Current

The subthreshold leakage current in a short-channel MOSFET can be characterized by the following expressions [48]:

$$I_{subthreshold} = \frac{\mu W C_{ox}}{L} V_T^2 e^{\frac{|V_{GS}| - |V_t|}{n V_T}} \left(1 - e^{\frac{-|V_{DS}|}{V_T}}\right), \tag{2.14}$$

$$V_T = \frac{kT}{q}, \tag{2.15}$$

where μ is the carrier mobility, W is the transistor width, C_{ox} is the oxide capacitance per unit area, V_T is the thermal voltage, V_{GS} is the gate-to-source voltage, V_t is the threshold voltage, n is the subthreshold swing coefficient, V_{DS} is the drain-to-source voltage, k is the Boltzmann constant (1.38×10^{-23} J/K), T is the absolute temperature (K), and q is the unit charge (1.6×10^{-19} C). The subthreshold swing coefficient (the ideality factor [40]) is the reciprocal of the rate of change in the channel surface potential as a function of the gate voltage [40], [100]. The subthreshold swing coefficient for a bulk MOSFET is [40]

$$n \cong 1 + \frac{\varepsilon_{Si} t_{ox}}{\varepsilon_{ox} t_{Si}}, \tag{2.16}$$

where ε_{Si} and ε_{ox} are the permittivity of silicon and gate oxide, respectively, t_{Si} is the thickness of the depletion layer of the substrate, and t_{ox} is the physical thickness of the gate oxide.

Assuming the body effect is approximately linear for low source-to-body voltages [37], the threshold voltage of a short-channel MOSFET is

$$V_t \cong V_{t_0} + \gamma V_{SB} - \eta V_{DS}, \tag{2.17}$$

where V_{t_0} is the zero body bias threshold voltage, γ is the body effect coefficient (assuming a linear approximation), and η is the DIBL coefficient.

Weak inversion is the most significant source of the total leakage current in current deep submicrometer CMOS technologies [46], [48]. As given by (2.14) and (2.17), the subthreshold leakage current is exponentially dependent on the junction temperature and the gate-to-source, drain-to-source, and threshold voltages. The exponential variation of the subthreshold leakage current of an n-type MOSFET with temperature for four different CMOS technology generations is illustrated in Figure 2.5. As shown in Figure 2.5, the subthreshold leakage current increases with technology scaling due to the lower threshold voltages and increasing short-channel and DIBL effects.

A commonly used parameter in characterizing the leakage behavior of deep submicrometer circuits is the subthreshold slope (also called the gate swing or the subthreshold swing [100]). The subthreshold slope (S) is the amount of variation required in the gate-to-source (V_{GS}) or threshold (V_t) voltage in order to vary the weak inversion current by one decade [36], [100]. As shown in Figure 2.2, the rate of change of the logarithm of the drain current (log (I_D)) with respect to the gate voltage is approximately linear in the subthreshold region. The subthreshold slope (mV/decade) can be evaluated by choosing two points in the subthreshold region of an I_D–V_{GS} curve such that the subthreshold leakage current changes by a factor of ten. From (2.14),

$$e^{\frac{|V_{GS1}| - |V_{GS2}|}{nV_T}} = 10, \tag{2.18}$$

$$S = |V_{GS1}| - |V_{GS2}| = nV_T \ln 10, \tag{2.19}$$

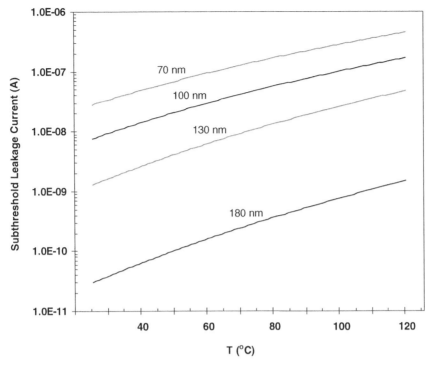

Figure 2.5   Variation of the subthreshold leakage current with junction temperature for four different CMOS technology generations ($W = 10W_{min}$ and $L = L_{min}$)

where V_{GS1} and V_{GS2} are the two gate-to-source voltages (in the subthreshold region of the I_D–V_{GS} curve) between which the subthreshold current varies by one decade.

Substituting S into (2.14),

$$I_{subthreshold} = \frac{\mu W C_{ox}}{L} V_T^2 e^{\frac{\ln 10(|V_{GS}|-|V_t|)}{S}} \left(1 - e^{\frac{-|V_{DS}|}{V_T}}\right),$$

(2.20)

$$I_0 = \frac{\mu W C_{ox}}{L} V_T^2 e^{\frac{|V_{GS}|\ln 10}{S}} \left(1 - e^{\frac{-|V_{DS}|}{V_T}}\right),$$

(2.21)

$$I_{subthreshold} = I_0 e^{\frac{-|V_t|\ln 10}{S}} = I_0 y,$$

(2.22)

$$y = e^{\frac{-|V_t|\ln 10}{S}} \Rightarrow \ln y = \frac{-|V_t|\ln 10}{S} \Rightarrow \frac{\ln y}{\ln 10} = \frac{-|V_t|}{S} \Rightarrow \log_{10} y = \frac{-|V_t|}{S},$$

(2.23)

$$y = 10^{\frac{-|V_t|}{S}},$$

(2.24)

$$I_{subthreshold} = I_0 10^{\frac{-|V_t|}{S}}.$$

(2.25)

Equations (2.14), (2.20), (2.22), and (2.25) are the commonly used expressions for subthreshold leakage current found in the literature. Equation (2.14) is a simplified version of the Berkeley short-channel insulated gate field effect transistor model (BSIM) [49]. Equation (2.25) captures the approximately linear relationship between the logarithmic leakage current and the threshold voltage of a MOSFET operating in the subthreshold region (see Figure 2.2). Alternatively, the following expression is often used to emphasize the logarithmic relationship between the gate-to-source and threshold voltages and the subthreshold leakage current:

$$I_{subthreshold} = I_0' 10^{\frac{(|V_{GS}|-|V_t|)}{s}},\tag{2.26}$$

$$I_0' = \frac{\mu W C_{ox}}{L} V_T^2 \left(1 - e^{\frac{-|V_{DS}|}{V_T}}\right).\tag{2.27}$$

Equation (2.14) is valid assuming $0 \le V_{GS} < V_t$ and $V_t < V_{GS} \le 0$ for NMOS and PMOS transistors, respectively. A more generic expression that characterizes subthreshold leakage current produced by NMOS and PMOS transistors also for small V_{GS} around zero for which $V_{GS} < 0$ and $V_{GS} > 0$, respectively, is

$$I_{subthreshold} = \frac{\mu W C_{ox}}{L} V_T^2 e^{\frac{-|V_{GS}-V_t|}{nV_T}} \left(1 - e^{\frac{-|V_{DS}|}{V_T}}\right).\tag{2.28}$$

A low subthreshold slope is desirable as the subthreshold leakage current decreases exponentially with reduced S. As given by (2.16) and (2.19), S can be reduced by lowering the thickness of the gate oxide and/or the doping concentration of the substrate (due to the increasing thickness of the depletion layer of the substrate). Another way to reduce S is to lower the junction temperature. As given by (2.16), if the depletion capacitance is assumed to be negligible as compared to the oxide capacitance ($C_{DEPLETION}/C_{ox} \approx 0$ and $n \approx 1$), the lower boundary of S is

$$S \ge V_T \ln 10.\tag{2.29}$$

As given by (2.29), a subthreshold slope of 60 mV/decade is a lower theoretical limit at room temperature for bulk silicon MOSFETs [84]. This minimum value of the subthreshold slope can be approached by fully depleted SOI devices [36], [73], [75]. Typical values of S vary between 80 mV/decade and 100 mV/decade at room temperature for bulk silicon MOSFETs fabricated in current CMOS technologies [51]–[55]. A comparison of the subthreshold slope, drain saturation current (I_{DSAT}), and off-state leakage current (I_{OFF}) of an NMOS transistor fabricated in different process technologies is listed in Table 2.1 [4], [51]–[55].

2.2.2 Gate Oxide Leakage Current

The gate insulator leakage current mechanisms in deeply scaled nanometer MOSFETs are reviewed in this section. Gate oxide leakage current increases with the scaling of the gate oxide thickness. Future scaling trends of gate oxide thickness and tunneling current are examined in Section 2.2.2.1. The various parameters that characterize gate oxide leakage current in nanometer MOSFETs are reviewed in Section 2.2.2.2. Alternative gate dielectric materials that are likely to replace SiO_2 as the gate insulator in the near future are discussed in Section 2.2.2.3.

Table 2.1 A Comparison of the Subthreshold Slope (S) and Leakage Current (I_{OFF}) of NMOS Transistors Fabricated in Different Technologies ($T = 25°C$)

Technology (μm)	L_{eff} (μm)	V_{DD} (V)	V_t (V)	I_{DSAT} (mA/μm)	I_{OFF} (nA/μm)	I_{DSAT}/I_{OFF}	S (mV/decade)
0.80	0.55	5.0	0.60	—	5.8×10^{-5}	—	86
0.60	0.35	3.3	0.58	—	1.5×10^{-4}	—	80
0.35	0.25	2.5	0.47	—	8.9×10^{-3}	—	80
0.25	0.15	1.8	0.43	—	24×10^{-3}	—	78
0.18	0.10	1.6	0.40	—	86×10^{-3}	—	85
0.18	0.10	1.5	0.29	1.04	3	347×10^3	90
0.13	0.06	1.4	0.30	1.14	10	114×10^3	85
(Dual-V_t) 0.13	0.06	1.4	0.27	1.30	100	130×10^2	85
0.10	0.05	1.2	0.34	0.95	30	316×10^2	87
0.09	0.05	1.2	—	1.26	40	315×10^2	85
(Dual-V_t) 0.09	0.05	1.2	—	1.45	400	3600	85

2.2.2.1 Effect of Technology Scaling on Gate Oxide Leakage

There are a number of challenges for continuing the scaling of MOSFET devices. A primary and most immediate challenge is imposed by continuing the scaling of the gate dielectric thickness. The scaling of the gate oxide thickness is crucial to enhancing the performances of MOSFETs [40], [51]–[56]. Reducing the thickness of the gate oxide (t_{ox}) increases the oxide capacitance per unit area, thereby enhancing the drain current of a MOSFET. The gate insulator in a MOSFET should be thin as compared to the device channel length in order for the gate to exert dominant control over the charge distribution in the channel as compared to the source and drain terminals of a device. An oxide thickness much smaller than the channel length reduces the short-channel effects (discussed in Section 2.2.1.1) [51]–[56]. Moreover, scaling the thickness of the gate oxide reduces the subthreshold slope, thereby lowering the subthreshold leakage current. The dielectric thickness is a few percent of the channel length in a typical MOSFET [56]. Future scaling trends of the gate oxide in relation to general CMOS technology scaling trends, extracted from the projections of the International Technology Roadmap for Semiconductors, are listed in Table 2.2 and shown in Figure 2.6 [56].

The quantum mechanical tunneling of carriers increases exponentially with decreasing insulator layer thickness [37], [40], [56]–[60]. T_{ox} is in the range of 12 Å to 16 Å in current

Table 2.2 Semiconductor Device Scaling Trends

Year of Production	1999	2002	2005	2008	2011	2014
Minimum feature size (nm)	180	130	100	70	50	35
Gate length (nm)	100	70	50	35	24	18
DRAM bits/chip	1G	3G	8G	24G	64G	192G
DRAM chip size (mm²)	400	460	530	630	710	860
Physical gate oxide Thickness (nm)	1.9–2.5	1.5–1.9	1.0–1.5	0.8–1.2	0.6–0.8	0.5–0.6
Dielectric constant of DRAM capacitor	22	50	250	700	1500	1500
Power supply (V)	1.5–1.8	1.2–1.5	0.9–1.2	0.6–0.9	0.5–0.6	0.5

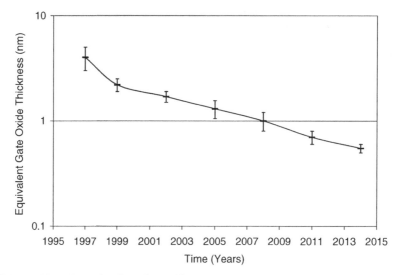

Figure 2.6   Scaling of gate insulator layer thickness. For each technology generation, the electrical gate oxide thicknesses span a range determined by the available or projected semiconductor technologies. Figure data: International Technology Roadmap of Semiconductors (ITRS) 1998–2001

CMOS technologies [162], [163]. Such a thin insulator layer can conduct a significant gate tunneling current despite scaling of the supply voltage. The SiO_2 gate oxide tunneling current density versus gate voltage for various insulator thicknesses of an NMOS device is shown in Figure 2.7 for a 100 nm CMOS technology [40]–[43].

Assume that the drain current of a MOSFET in a 100 nm technology generation is approximately 1 mA/μm [56]. Also assume that the gate tunneling current is limited to

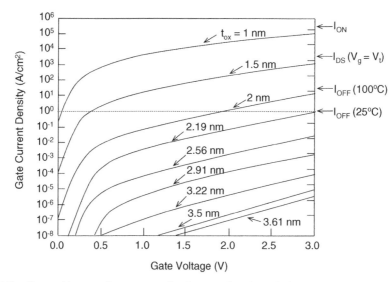

Figure 2.7   Gate oxide tunneling current density as a function of the gate voltage for various gate oxide (SiO_2) thicknesses assuming a 100 nm CMOS technology [40]–[43]

1% of the drain current in order to not significantly degrade the gain of the devices [56]. These assumptions constrain the acceptable maximum gate leakage current density to 10^4 A/cm^2. As shown in Figure 2.7 and assuming a supply voltage as listed in Table 2.2, a gate oxide thickness in the range of 1.0 nm to 1.5 nm results in acceptable gate leakage without significantly degrading the gain of the devices in a 100 nm technology generation. These high levels of gate oxide tunneling current, however, significantly degrade the device reliability and increase the power consumption of deeply scaled CMOS circuits [37], [40], [56], [58], [59].

The direct tunneling currents that continuously flow in thin oxide layers create long-term reliability concerns [56]. The oxide defect density gradually increases due to the oxide leakage current. Eventually, the gate oxide destructively breaks down (via a short circuit) [56], [58], [59]. As described in [58], the hard breakdown (failure) of the gate insulator is highly dependent on the thickness of the insulator and the gate voltage. As the oxide tunneling current increases exponentially with decreasing oxide thickness and as scaling of the supply voltage slows down due to the difficulty to scale the threshold voltages, the time to breakdown will decrease with technology scaling [56], [58], [59].

The variation of the gate oxide and subthreshold leakage currents with supply voltage and temperature in a 45 nm CMOS technology is shown in Figure 2.8. The subthreshold leakage current increases exponentially with temperature while the gate oxide leakage current displays a weaker dependence on temperature. The relative contribution of the subthreshold and gate oxide leakage currents to the total leakage power consumption, therefore, changes with die temperature. At 110°C, the subthreshold leakage current produced by a low-V_t NMOS transistor is 6.8 times greater than the gate oxide leakage current at the nominal supply voltage ($V_{DD} = 0.8$ V), as shown in Figure 2.8. Alternatively, at room temperature, I_{gate} is 2.5 times

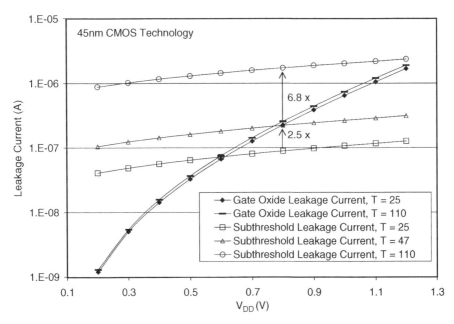

Figure 2.8 Comparison of subthreshold and gate oxide leakage currents produced by an NMOS transistor for various supply voltages at three temperatures. $I_{subthreshold}$: $V_{GS} = 0$ and $V_{DS} = V_{DD}$. I_{gate}: $V_{GS} = V_{GD} = V_{GB} = V_{DD}$

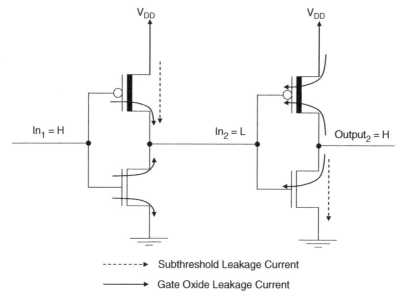

-------► Subthreshold Leakage Current

———► Gate Oxide Leakage Current

Figure 2.9   Leakage current paths in a CMOS circuit

higher than the subthreshold leakage current. Most of the power consumption of an idle circuit can be caused by gate dielectric tunneling current, particularly at low die temperatures during long idle periods. Gate oxide leakage current is expected to become a primary source of power consumption in nanometer CMOS circuits in the near future.

2.2.2.2 Characterization of Gate Oxide Leakage Current

A schematic of the subthreshold and gate oxide leakage current paths in two cascaded CMOS inverters is shown in Figure 2.9. When the gate of an NMOS device is positively biased, an inversion layer is formed underneath the gate. The electrons in the inverted channel can tunnel to the positively biased poly gate, producing a gate oxide leakage current. Similarly, when the gate of a PMOS transistor is negatively biased, the holes in the inverted channel can tunnel to the poly gate, producing a hole tunneling leakage current.

The two gate tunneling mechanisms in MOSFETs are Fowler–Nordheim tunneling through the oxide layer conduction band and direct tunneling through the forbidden energy gap of the gate insulator [37], [178], [179]. Fowler–Nordheim tunneling is observed at unusually high oxide voltages (oxide voltage V_{ox} > barrier height ϕ_b) due to electrons tunneling into the conduction band of the gate insulator [178], [179]. Such high oxide voltages are typically not encountered during the normal operation of a CMOS circuit [37]. Alternatively, direct tunneling occurs under normal bias conditions ($V_{ox} < \phi_b$) of a MOSFET with a very thin oxide layer, as the electrons or holes in the inverted silicon surface tunnel directly through the forbidden energy gap of the ultra-thin gate insulator (typically observed for t_{ox} less than about 4 nm) [179]. The focus of this section is on the characterization of direct tunneling current through thin gate insulators.

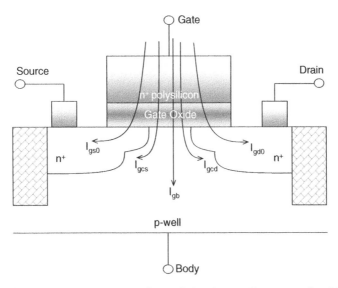

Figure 2.10 Different components of gate dielectric tunneling current in a MOSFET

The gate tunneling current is composed of several components as illustrated in Figures 2.10 and 2.11 [39], [57]. I_{gb} is the gate-to-substrate leakage current produced by electron direct tunneling from the valence band in both NMOS and PMOS devices [177]. I_{gs0} and I_{gd0} are the leakage currents through the gate-to-source and gate-to-drain overlap regions, respectively. I_{gc} is the gate-to-channel tunneling current during operation in the inversion region. A portion of I_{gc} is collected by the source while the remaining portion of I_{gc} is collected by the drain. The highest gate oxide leakage current is observed when a transistor operates in the active region with the maximum voltage difference across the gate-to-source and the gate-to-drain terminals, as illustrated in Figure 2.11(a). Alternatively, the highest subthreshold leakage current is observed when a cut-off transistor is biased with the maximum voltage difference between the source and drain terminals, as depicted in Figure 2.11(b).

Direct tunneling current in an MOS device depends on the voltage across the gate dielectric, the thickness of the dielectric, the tunneling barrier height, the effective mass of the carriers, and the number of free carriers available for tunneling on the MOS electrodes [56], [177].

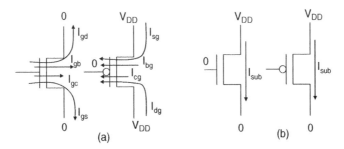

Figure 2.11 Bias conditions that produce maximum gate oxide and subthreshold leakage current in NMOS and PMOS transistors. (a) Maximum gate oxide leakage current state. (b) Maximum subthreshold leakage current state

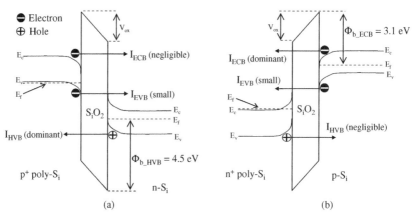

Figure 2.12   The primary mechanisms of gate dielectric direct tunneling current in PMOS and NMOS transistors. (a) A PMOS transistor in inversion. (b) An NMOS transistor in inversion

Different mechanisms of gate dielectric direct tunneling are illustrated by the MOSFET energy band diagrams in Figure 2.12. The primary sources of gate dielectric direct tunneling leakage current in a MOSFET are electron tunneling from the conduction band (ECB), electron tunneling from the valence band (EVB), and hole tunneling from the valence band (HVB) [39], [57].

A semi-empirical gate oxide tunneling model is presented in [175]–[177] for estimating the direct tunneling leakage current produced by MOSFETs. This model accurately estimates the electron tunneling current from the conduction band, the electron tunneling current from the valence band, and the hole tunneling current from the valence band in NMOS and PMOS transistors under typical gate bias conditions. Assuming the voltage across the gate oxide is less than the barrier height of carriers ($V_{ox} < \phi_{b\_i}$), the direct tunneling current density (A/cm^2) is

$$J_i = D_i A_i \frac{1}{\varepsilon_{ox}} e^{\frac{B_i \left[1 - (1 - |V_{ox}|/\phi_{b\_i})^{\frac{3}{2}}\right]}{|E_{ox}|}}, \qquad (2.30)$$

$$A_i = \frac{q^3}{8\pi h \phi_{b\_i}}, \qquad (2.31)$$

$$B_i = \frac{-8\pi \sqrt{2m_{ox\_i}} \left(\phi_{b\_i}^{\frac{3}{2}}\right)}{3hq}, \qquad (2.32)$$

$$E_{ox} = \frac{V_{ox}}{t_{ox}}, \qquad (2.33)$$

where i is the index to distinguish the different mechanisms of gate oxide tunneling (ECB, HVB, and EVB), D_i is a semi-empirical correction function determined by curve fitting, ε_{ox} is the dielectric constant of gate oxide, V_{ox} is the voltage difference across the gate oxide (see Figure 2.11), ϕ_{b_i} is the carrier tunneling barrier height, E_{ox} is the electric field in the gate oxide, t_{ox} is the physical thickness of gate oxide, q is the electronic charge, h is the Planck constant, and m_{ox_i} is the effective mass of tunneling carriers in the gate oxide. ϕ_{b_i} is 3.1 eV for electrons tunneling from the conduction band (ϕ_{b_ECB}), 4.2 eV for electrons tunneling from the valence band (ϕ_{b_EVB}), and 4.5 eV for holes tunneling from the valence band (ϕ_{b_HVB}) assuming a polysilicon gate [175]–[177]. Typical empirical values of m_{ox_i} presented in [175]–[177] are $0.4m_0$ for ECB, $0.3m_0$ for EVB, and $0.32m_0$ for HVB (m_0 is the mass of a free electron).

As described previously, the dominant mechanism of tunneling in MOSFETs is determined by the relative value of the oxide voltage as compared to the tunneling barrier height. Direct tunneling of carriers through the forbidden energy gap of the gate insulator is the dominant source of gate oxide leakage current for $V_{ox} < \phi_{b\_i}$. As given by (2.30), tunneling current density is strongly dependent on the oxide voltage. The oxide voltage is

$$V_{ox} = V_{ge} - \phi_s - V_{FB}, \qquad (2.34)$$

$$V_{ge} = V_g - V_{poly}, \qquad (2.35)$$

$$\phi_s = \left[\frac{\gamma}{2} \left(-1 + \sqrt{1 + \frac{4(V_g - V_{ge} - V_{FB})}{\gamma^2}} \right) \right]^2, \qquad (2.36)$$

$$\gamma = \frac{\sqrt{2\varepsilon_{Si} q N_{ch}}}{C_{ox}}, \qquad (2.37)$$

where V_{ge} is the effective gate voltage considering the voltage drop across the polysilicon depletion region (V_{poly}), ϕ_s is the band bending at the surface, V_{FB} is the flat band voltage, V_g is the gate voltage, γ is the body effect coefficient, ε_{Si} is the dielectric constant of silicon, N_{ch} is the channel doping concentration, and C_{ox} is the gate oxide capacitance per unit area.

The correction function is

$$D_i = N \frac{V_g}{t_{ox}} e^{\frac{20}{\phi_{b\_i}} \left(\frac{|V_{ox}| - \phi_{b\_i}}{\phi_{b0}} + 1 \right)^{\alpha_i}} \left(1 - \frac{|V_{ox}|}{\phi_{b0}} \right), \qquad (2.38)$$

where N is the density of carriers ($\mu C/cm^2$) in the inversion or accumulation layer of the carrier injecting electrode, ϕ_{b0} is the Si/SiO$_2$ barrier height, and α_i is a fitting parameter that depends on the tunneling mechanism. ϕ_{b0} is 3.1 eV for electrons and 4.5 eV for holes [175]–[177]. α_i accounts for the second order non-ideal characteristics such as the finite density of carriers and energy states in the semiconductor and the dependence of the effective mass of carriers on the energy state. Typical empirical values of α_i for ECB, EVB, and HVB are 0.6, 1, and 0.4, respectively [177]. ECB and HVB are the dominant mechanisms of tunneling in NMOS and PMOS transistors, respectively, operating in the inversion region as illustrated in Figure 2.12. The carrier density for ECB or HVB is

$$N = \frac{\varepsilon_{ox}}{t_{ox}} \left[n_{inv} V_T \ln \left(1 + e^{\frac{V_{ge} - V_t}{n_{inv} V_T}} \right) + n_{acc} V_T \ln \left(1 + e^{\frac{V_{FB} - V_g}{n_{acc} V_T}} \right) \right], \qquad (2.39)$$

$$n_{inv} = \frac{S}{V_T}, \qquad (2.40)$$

$$V_T = \frac{kT}{q}, \qquad (2.41)$$

where n_{inv} and n_{acc} are swing parameters, S is the subthreshold slope, V_T is the thermal voltage, k is the Boltzmann constant, T is the absolute temperature, and V_t is the threshold voltage. Substituting (2.40) and 1 (default value for an NMOS transistor [177]) for n_{inv} and n_{acc}, respectively, in (2.39)

$$N = \frac{\varepsilon_{ox}}{t_{ox}}\left[S\ln\left(1 + e^{\frac{V_{ge}-V_t}{S}}\right) + V_T\ln\left(1 + e^{\frac{V_{FB}-V_g}{V_T}}\right)\right]. \tag{2.42}$$

N provides a measure of the number of free carriers (mobile charge density) available for tunneling. As given by (2.42), N is inversely proportional to the oxide thickness. Scaling of the oxide thickness increases the correction function because of both the increase in the number of carriers and the increase in the inverse proportionality term (V_g/t_{ox}) as given by (2.38). Furthermore, the exponential term in (2.30) increases due to the enhanced electric field within the gate oxide as the oxide thickness is scaled for a given gate bias voltage. The gate oxide leakage current, therefore, significantly increases with the scaling of gate insulator thickness in each new technology generation. Equation (2.42) can also be used to characterize the effect of temperature on gate oxide leakage current. The subthreshold slope S is increased while the threshold voltage V_t is reduced with increased temperature. The number of free carriers available for tunneling therefore increases with increased die temperature, thereby enhancing the gate oxide leakage current density as given by (2.42) and (2.30).

The first term in the parentheses in (2.42) models the carrier density due to inversion. The onset of effective inversion occurs when V_{ge} exceeds V_t. The first term in (2.42) is the dominant term when a transistor is biased for inversion ($V_g > 0$ for NMOS transistors and $V_g < 0$ for PMOS transistors). Alternatively, the second term in the parentheses in (2.42) models the free carrier density due to accumulation. The onset of effective accumulation occurs when $V_g < V_{FB}$ and $V_g > V_{FB}$ for NMOS and PMOS transistors, respectively. When a device has a strongly inverted channel or is in accumulation, carrier density becomes a linear function of ($V_{ge} - V_t$) or ($V_{FB} - V_g$), respectively. The second term in (2.42) is dominant when a transistor is biased into accumulation ($V_g < V_{FB} < 0$ for a NMOSFET and $V_g > V_{FB} > 0$ for a PMOSFET). In the subthreshold and subflatband regions ($|V_{ge}| < |V_t|$ and $|V_g| < |V_{FB}|$, respectively), the carrier density decreases exponentially with reduced gate voltage. The tunneling current is reduced due to the decreased number of free carriers in the subthreshold and subflatband regions.

In an NMOS transistor, electron tunneling from the conduction band is the dominant mechanism of gate leakage in both the inversion and accumulation regions of operation. Alternatively, for small positive gate voltages which cannot induce a conducting channel, electron direct tunneling from the valence band to the body becomes the dominant leakage mechanism. The hole concentration in the n^+ polysilicon gate is very small. The gate-to-substrate hole tunneling current I_{gb} is therefore negligible as compared to I_{gc} in an active (inverted) NMOS transistor.

Electron tunneling from the conduction band is the dominant leakage mechanism in a PMOS device operating in the accumulation region. Alternatively, for a PMOS transistor operating in inversion, hole tunneling from the valence band is the dominant mechanism of gate oxide leakage, as illustrated in Figure 2.12. Electron concentration in the p^+ polysilicon gate is very small in a PMOS transistor. Substrate-to-gate electron tunneling current I_{bg} is therefore negligible as compared to I_{cg} in an active PMOS transistor. For high negative voltages, electron direct tunneling from the valence band to the body becomes the dominant leakage mechanism. The primary tunneling current mechanisms that generate the gate

Table 2.3 The Dominant Mechanisms of Gate Oxide Tunneling Current for Different Regions of Operation of a MOSFET

Current component		I_{gc}	I_{gb}	I_{gb}
Region of operation		Inversion	$V_g > 0$	$V_g < 0$
Type of transistor	PMOS	HVB	ECB	EVB
	NMOS	ECB	EVB	ECB

leakage current components for different regions of operation of n-channel and p-channel MOSFETs are listed in Table 2.3 [57].

In CMOS technologies with a SiO_2 gate insulator, due to the higher tunneling barrier for holes ($\phi_{b\_HVB} > \phi_{b\_ECB}$ as shown in Figure 2.12), the hole tunneling current is more than an order of magnitude smaller than the electron tunneling current for the same physical dimensions and bias conditions of the PMOS and NMOS transistors, as shown in Figure 2.13. The gate oxide leakage current of an SiO_2-based CMOS circuit is, therefore, typically dominated by the gate oxide leakage current produced by the active (inverted) NMOS transistors. In CMOS technologies utilizing gate dielectric materials other than SiO_2, however, hole tunneling current produced by PMOS transistors can be the dominant source of gate oxide leakage current. For example, in a CMOS technology with a SiN_3 gate insulator, the barrier height for holes is smaller than the barrier height for electrons ($\phi_{b\_HVB} = 1.9$ eV $< \phi_{b\_ECB} = 2.1$ eV) [176]. In SiN_3-based CMOS technologies, therefore, the gate oxide leakage current of a CMOS circuit is dominated by the tunneling currents produced by the active (inverted) PMOS transistors through HVB.

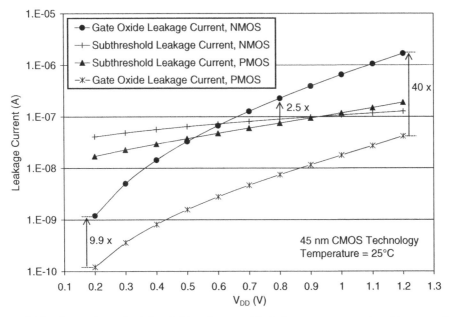

Figure 2.13 Comparison of subthreshold and gate oxide leakage currents produced by same-sized NMOS and PMOS transistors for various supply voltages. $I_{subthreshold}$: $V_{GS} = 0$ and $V_{DS} = V_{DD}$. I_{gate}: $V_{GS} = V_{GD} = V_{GB} = V_{DD}$

2.2.2.3 Alternative Gate Dielectric Materials

Silicon dioxide (SiO_2) has been the material of choice as the gate oxide insulator for the past three decades. SiO_2 is relatively easy to grow on silicon, forming an abrupt interface with near ideal electrical characteristics [37], [40], [56]. The trap and fixed charge densities at the Si–SiO_2 interface are typically less than one surface defect in 10^5 surface silicon atoms, forming a nearly ideal interface between Si and SiO_2 [56].

The dielectric thicknesses projected in Table 2.2 will soon become unrealizable if SiO_2 is maintained as the insulator material. New materials with higher dielectric constant (high K) as compared to SiO_2 are required as the gate insulator in order to keep the gate oxide tunneling current under control [55], [56]. A material with a higher dielectric constant will permit a physically thicker dielectric layer while offering an equivalent unit oxide capacitance that corresponds to the values listed in Table 2.2. A thicker and higher K gate insulator material can significantly reduce oxide tunneling leakage current while increasing the time to break down the oxide, thereby enhancing both energy efficiency and device reliability [37], [56]. Some strong candidates that are likely to replace SiO_2 as the gate oxide material in the near future are shown in Figure 2.14 [17].

As discussed previously, gate oxide tunneling current depends not only on the thickness of the insulator but also on the effective mass of the carriers and the barrier height of the insulator [56]. Although employing a high K dielectric provides the opportunity to increase the thickness of the insulator while simultaneously enhancing capacitive coupling between the gate and the channel, a higher K material also typically decreases the bandgap of the

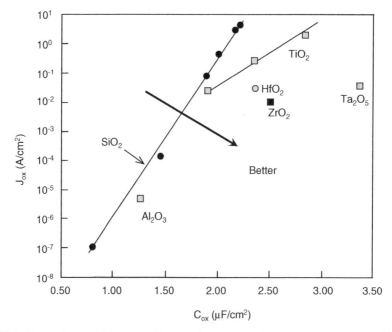

Figure 2.14 Comparison of the gate oxide capacitance per unit area versus the gate oxide leakage current density of various insulators [17] for aluminum oxide (Al_2O_3), hafnium dioxide (HfO_2), silicon dioxide (SiO_2), tantalum pentoxide (Ta_2O_5), titanium dioxide (TiO_2) and zirconium dioxide (ZrO_2)

insulator. A large bandgap is desirable as the barrier height typically scales with the bandgap voltage [40], [56]. A high K dielectric should, therefore, be carefully chosen as a thicker insulator does not always result in a significantly lower tunneling current.

Another issue with replacing SiO_2 with a high K material is the possible degradation of the interface between the silicon and the new dielectric material. Provided that the new high K material cannot offer a low defect density comparable to the Si–SiO_2 interface, the potential improvement in the transistor current due to enhanced capacitive coupling between the gate and the channel can be offset by degraded carrier mobility at the surface [56]. Increasing the dielectric constant above a certain limit degrades circuit performance due to a lower surface mobility, provided that similar subthreshold leakage current characteristics comparable to a standard SiO_2 MOSFET are maintained [60]. It has also been shown in [60] that employing a thicker high K dielectric can increase short-channel effects, thereby increasing the subthreshold leakage current and threshold voltage roll-off due to increasing fringing electric fields from the gate to the source and drain.

2.3 SHORT-CIRCUIT POWER

In static CMOS circuits, there is a time period during the transition of the input signals when both the pull-up and pull-down network transistors are simultaneously on, thereby forming a DC current path between the power supply and ground. The DC current conducted by a CMOS circuit during an input signal transient (due to non-zero rise and fall times of the input signals) is called the short-circuit current [9], [36], [61]. The short-circuit current ($I_{short-circuit}$) is temporarily observed during the input signal transition, $V_{tn} \leq V_{in} \leq V_{DD} + V_{tp}$, as illustrated in Figure 2.15.

Short-circuit current is a function of the output load and the rise and fall times of the input and output signals. The variation of the short-circuit current with the load capacitance is illustrated in Figure 2.16. The short-circuit current can be significant when the rise and fall times of the input signals are significantly larger than the output rise and fall times as the short-circuit current path will exist for a longer period of time [36], [61].

The output signal waveforms produced by an inverter driving different loads are shown in Figure 2.17. For a given input signal, the output transition is faster as the output capacitance (C_L) is reduced. The short-circuit current for $C_L = 0$ is, therefore, the maximum short-circuit current that can be produced by a CMOS circuit. As the output capacitance is increased, the output rise and fall times also increase. For a high load capacitance, which translates into a slow output transition, the pull-up and pull-down network transistors experience a negligible voltage difference between the source and drain terminals during the input rise and fall times, respectively. The window of opportunity for the pull-up (or pull-down) network transistors to produce short-circuit current, therefore, narrows with increasing load capacitance. Although the pull-up and pull-down network transistors simultaneously operate in the active region, a small amount of short-circuit current will be conducted during the input transition for a high output capacitance as illustrated in Figure 2.16. Theoretically, the short-circuit current approaches zero as the load capacitance approaches infinity.

As discussed in [9], [36], and [61], the short-circuit power typically contributes to less than 10% of the total power consumed in a CMOS circuit provided that the input slew rate is higher than the output slew rate. The short-circuit power consumption can dominate the

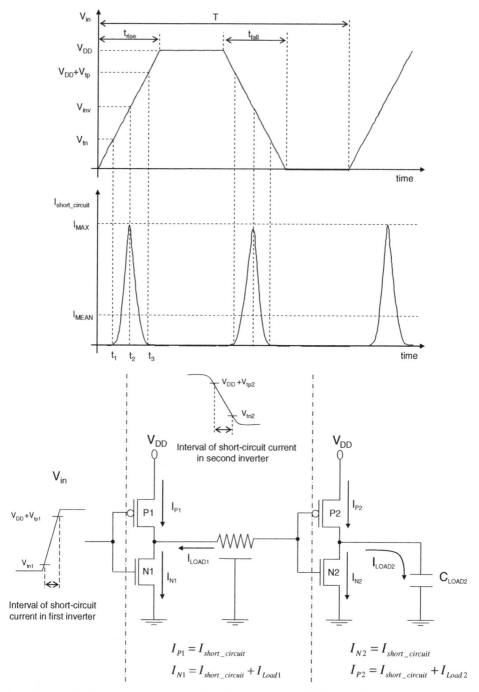

Figure 2.15   The short-circuit currents produced by two cascaded inverters. V_{tn} = threshold voltage of an NMOS transistor. V_{tp} = threshold voltage of a PMOS transistor. V_{inv} = switching threshold voltage of an inverter

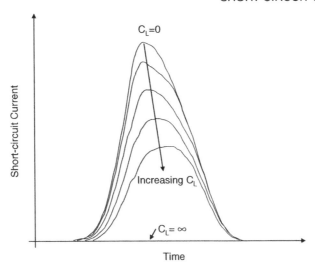

Figure 2.16 The variation of short-circuit current with the output load capacitance. The short-circuit current is a maximum at no load. Alternatively, the short-circuit current is a minimum at an infinite load

total power consumption of a CMOS circuit if the output is lightly loaded as illustrated in Figure 2.18. Similarly, the short-circuit power consumption can be as high as the dynamic switching power consumption if the input signal rise and fall times are unusually long as illustrated in Figure 2.19.

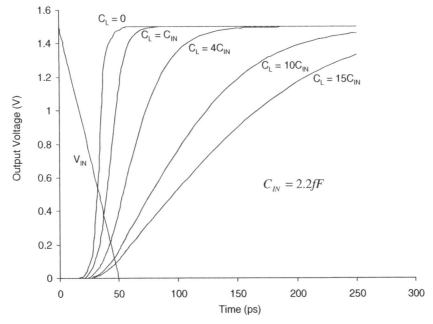

Figure 2.17 The output waveforms produced by a CMOS circuit for different output load capacitances. The window of opportunity to produce short-circuit current narrows as the load capacitance is increased

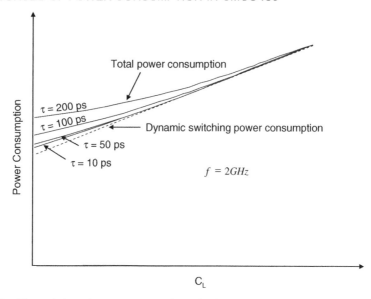

Figure 2.18   The variation of power consumption with input rise and fall times and load capacitance

As discussed in [62], the contribution of the short-circuit power to the total power consumption is expected to be smaller with technology scaling due to the increasing threshold to supply voltage ratio (V_t/V_{DD}). The short-circuit current can be effectively eliminated by lowering the supply voltage below the sum of the threshold voltages of the PMOS and NMOS transistors, $V_{DD} < V_{tn} + |V_{tp}|$ [36]. A CMOS circuit technology that does

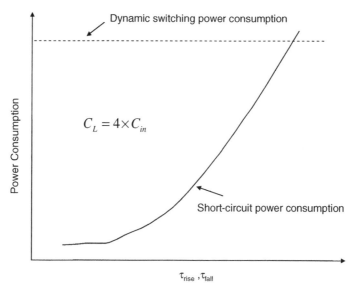

Figure 2.19   The dependence of the dynamic switching and short-circuit power on the input rise and fall times

not suffer from short-circuit current is ultra-low power subthreshold CMOS [99]. The transistors in a subthreshold logic circuit operate in the weak inversion region and do not consume any short-circuit power.

2.4 STATIC DC POWER

The change of the dominant IC technology from NMOS to CMOS in the early 1980s has diminished the issue of static DC power. CMOS circuits do not consume any static DC power (excluding leakage power) as long as the signal voltage at the internal nodes swings full rail between V_{DD} and ground. Non-full rail voltage levels, however, are sometimes encountered in CMOS circuits due to the employment of low signal swing circuitry such as NMOS pass gates [9] and low swing interconnect signaling techniques [32]. Non-full rail voltage levels can also be observed at the interfaces between different ICs or circuit styles (as in systems on-chip, SoC) where different circuits often operate at different voltage levels. When a CMOS circuit supplied by full rail power and ground supplies is driven by a low swing input signal, static power is dissipated as the transistors in both the pull-up and pull-down networks are simultaneously turned on. A CMOS inverter driven by a low swing signal is shown in Figure 2.20. The second stage gate in Figure 2.20 behaves as a voltage divider (rather than as an inverter), consuming static DC power and degrading the voltage swing at node$_2$.

In ICs with multiple supply voltages, energy-efficient and full swing signal transfer among regions operating at different voltage levels requires specialized voltage interface circuits. Voltage interface circuits are discussed in Chapter 8.

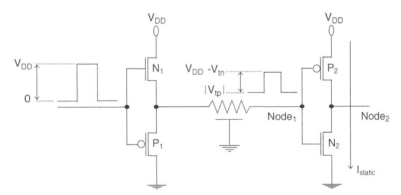

Figure 2.20 Static DC current in a full voltage rail CMOS inverter driven by a low voltage swing signal

3 Supply and Threshold Voltage Scaling Techniques

Supply voltage scaling is an essential step in the technology scaling process. Two primary reasons for scaling the supply voltage are to maintain the power density of an IC below a limit dictated by available cost-effective cooling techniques and to guarantee the long-term reliability of the devices fabricated in a scaled semiconductor technology.

Provided that the supply voltage is not scaled together with the vertical and lateral dimensions of the devices, the electric field across the terminals of the MOSFETs increase, degrading the reliability and changing the electrical characteristics of the devices. The electric field between the source and drain and across the source-to-body and drain-to-body junctions must be maintained below a certain level to lessen any short-channel effects in a scaled CMOS technology. The electric field across the gate oxide must also be limited to maintain high carrier mobility in the channel region [52], [54] and to lower tunneling-based leakage currents through the scaled gate insulator [51], [52], [54]. The gate oxide leakage currents can significantly increase the static power dissipation while decreasing the transconductance and gate oxide failure time of the devices [40], [56], [58], [59].

Dynamic switching energy is the dominant component of the total energy consumed by an IC in current CMOS technologies. The dynamic switching energy is proportional to the square of the supply voltage in a full voltage swing CMOS circuit. Moreover, the leakage and short-circuit energy components also depend superlinearly on the supply voltage. Reducing the supply voltage, therefore, is an effective way to lower the power consumption. The variation of the total power consumption of a 0.18 μm-based CMOS ring oscillator with supply voltage is shown in Figure 3.1.

The (high-to-low or low-to-high) propagation delay through a CMOS gate can be approximated by [9], [91]

$$T_d \cong \frac{C_L V_{DD}}{I} = \frac{L_{eff} C_L V_{DD}}{WB(V_{DD} - V_t)^n}, \tag{3.1}$$

$$T_d \propto \frac{V_{DD}}{(V_{DD} - V_t)^n}, \tag{3.2}$$

Multi-Voltage CMOS Circuit Design V. Kursun and E. Friedman
© 2006 John Wiley & Sons, Ltd

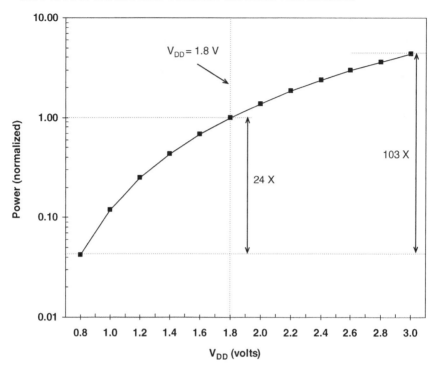

Figure 3.1   Normalized power consumption versus supply voltage (V_{DD}) of a 19 stage ring oscillator (for a 0.18 μm CMOS technology)

where C_L is the load capacitance, V_{DD} is the supply voltage, I is the drain current in the saturation region, W is the effective transistor width, and L_{eff} is the effective transistor length. B and n are technology-related parameters that determine the drain current characteristics of a deep submicrometer MOSFET operating in the saturation region [91]. The value of n varies between one and two depending upon the MOSFET fabrication technology. For a long-channel device, n is two. For a short-channel device, n is typically much less than two due to velocity saturation.

Combining (2.11) and (3.2) and assuming a full voltage swing circuit, the relationship between the dynamic switching power consumption and the power supply and threshold voltages is

$$P_d \propto V_{DD}(V_{DD} - V_t)^n. \tag{3.3}$$

Note that the relationship between the delay and the supply voltage is nonlinear. As given by (3.2), the delay of a CMOS circuit increases with reduced supply voltage. The variation of the delay of a CMOS ring oscillator with supply voltage is shown in Figure 3.2. Lowering the supply voltage (while maintaining the same threshold voltages) reduces both the energy consumed by the parasitic impedances (due to the lower amount of energy stored in the parasitic capacitors) and the maximum operating frequency. The reduction in power consumption by supply voltage scaling is, therefore, more than quadratic as given by (3.3) and as shown in Figure 3.1.

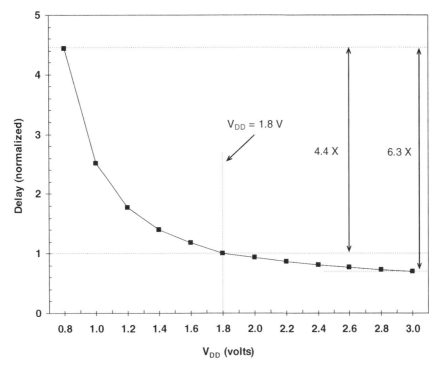

Figure 3.2   Normalized delay versus supply voltage (V_{DD}) of a 19 stage ring oscillator (for a 0.18 μm CMOS technology)

The primary reasons for the reluctance to move to lower supply voltages have been the speed penalty and compatibility with commercial power supplies. Until recently, the preferred value of supply voltage has typically been determined by device reliability requirements rather than by power dissipation concerns. Tradeoffs and priorities, however, have shifted as the power density of high performance ICs approach 100 W/cm^2 (see Figure 1.8). Available low cost cooling solutions are ineffective at such high power densities. Moreover, satisfying the market demand for enhanced performance and functionality in portable applications has become increasingly challenging due to lagging improvements in battery technology and cooling solutions. There is, therefore, a necessity for optimizing high performance ICs not only for higher speed and reliability but also for lower power dissipation. Supply voltage scaling is expected to continue into the foreseeable future, as scaling is the most effective technique for reducing power consumption in CMOS ICs.

Several techniques have been proposed for exploiting the more than quadratic reduction in power consumption by lowering the supply voltage while compensating for the speed degradation when operating at a lower supply voltage. At the architectural level, an effective way to maintain circuit performance while lowering the supply voltage is to utilize parallel (or pipelined) architectures. As discussed in [36], employing parallel circuits (each parallel circuit has a similar function) permits the clock speed requirement per circuit block to be reduced in order to execute a specific task with a target latency. Parallel circuit blocks can operate at a lower supply voltage at reduced speed while achieving overall circuit throughput objectives [36]. The primary disadvantage of this technique, however, is the significant area and power overhead due to the parallel replication of the circuitry [29].

At the circuit level, an effective way to lower power consumption without degrading performance is to dynamically adjust the supply voltage as the workload varies with time. Dynamic voltage scaling techniques are discussed in Section 3.1. Another technique for minimizing the deleterious effects of supply voltage scaling is to lower the supply voltage of only those circuits along the non-critical delay paths while maintaining a higher supply voltage on the speed-critical paths [28]. Multiple supply voltage circuit techniques that exploit differences in signal propagation delays along different delay paths by selectively scaling the local supply voltages are reviewed in Section 3.2.

The most widely employed technique for enhancing the performance of a circuit at a reduced supply voltage is to scale the threshold voltages. Lowering the threshold voltages enhances the gate overdrive ($|V_{GS}|-|V_t|$) of the transistors, thereby reducing the propagation delay of the circuits. The threshold voltage scaling technique is discussed in Section 3.3.

Threshold voltage scaling not only enhances the speed but also increases subthreshold leakage currents, short-channel effects, and die-to-die and within-die parameter variations. A promising circuit technique aimed at lowering deleterious side effects caused by supply and threshold voltage scaling is the use of multiple supplies and threshold voltages. Circuit techniques based on multiple supplies and threshold voltages are reviewed in Section 3.4.

Dynamic supply and threshold voltage scaling techniques combine the desirable characteristics of dynamic supply voltage scaling (in order to lower the power consumption) and dynamic threshold voltage scaling (in order to lower subthreshold leakage current and die-to-die and within-die variations of electrical characteristics). Dynamic supply and threshold voltage scaling techniques are discussed in Section 3.5.

Partitioning an IC into multiple circuit blocks with individually optimized supply voltages and clock frequencies can significantly enhance the effectiveness of dynamic voltage and frequency scaling techniques. In addition to multiple clock domains that operate in a typical globally asynchronous, locally synchronous (GALS) circuit, different circuit blocks operating at different supply voltages exist in a multiple voltage and clock domain IC. In addition to lowering the power consumption, a multiple voltage and clock domain circuit can also significantly reduce the complexity of the on-chip synchronization circuitry. Multiple supply and clock domain ICs are described in Section 3.6. A summary of the various supply and threshold voltage scaling techniques presented in this chapter is provided in Section 3.7.

3.1 DYNAMIC SUPPLY VOLTAGE SCALING

The computational load in a microprocessor system varies with time [68]–[71]. Applications in a typical microprocessor system tend to have peak performance requirements followed by idle periods, as shown in Figure 3.3. During active computation, the microprocessor performance required to execute a task varies as a function of the workload. An operation is a basic unit of computation [68]. Another commonly used basic unit of computation is an instruction. The utilization of a microprocessor can be evaluated in terms of the number of operations or the number of instructions required to complete the processing of a task within a specified time frame. A metric called *throughput* is often used as a measure of the utilization of a microprocessor system. The throughput is the

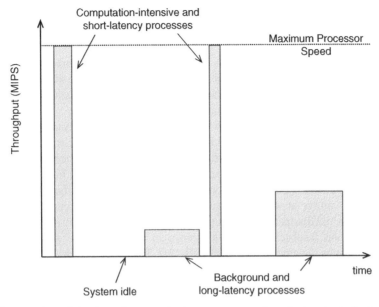

Figure 3.3   Variation of the throughput required to execute certain tasks in a typical microprocessor system [68]

number of operations (or the number of instructions) performed over a unit period of time [68],

$$Throughput = \frac{Number\ of\ Operations}{Unit\ Time}.$$ (3.4)

The throughput is typically described in terms of either millions of operations per second (MOPS) or millions of instructions per second (MIPS).

Computation-intensive and short-latency tasks (e.g., audio and video decompression, speech and image recognition) typically utilize the maximum throughput of a microprocessor [68]. Alternatively, low speed and long-latency tasks (e.g., playing music, periodic system checks, and backup) require only a fraction of the maximum throughput offered by a microprocessor. Executing the long-latency tasks faster than the required throughput has no observable benefit for the user [68], [70]. Moreover, as shown in Figure 3.3, there are frequent idle periods during which no active computation is made by a computer system (zero throughput requirement). Maintaining the full computational capacity of a processor during such idle periods wastes significant amounts of energy despite the zero throughput requirement. Particularly in portable applications such as cellular phones, the duration of these idle periods is typically significantly longer than the duration of the active periods [83]. Lowering the energy consumption of an IC during these long idle periods is therefore critical for extending the limited lifetime of a battery.

The dynamic voltage scaling (DVS) circuit technique exploits variations in the computational workload by dynamically modifying the supply voltage and clock frequency of a microprocessor system [68]–[70]. The primary objective of the DVS circuit technique is to provide high throughput during the execution of only the computation-intensive tasks while saving energy during the rest of the time by lowering the supply voltage and operating speed of a microprocessor.

The clock frequency of a DVS microprocessor is controlled by software. Only the operating system (OS) has the necessary information that characterizes all of the active tasks. The OS, therefore, controls the clock frequency by determining the minimum clock frequency required to complete a task within a specific time period.

Since the required clock frequency of a microprocessor varies, the supply voltage should also vary to minimize the energy consumption while guaranteeing the operation of the microprocessor circuitry at the revised clock frequency requested by the OS. The software is not aware of the minimum supply voltage required for a microprocessor to operate at a desired clock frequency. Translating a desired clock frequency (requested by the OS) to a particular minimum supply voltage at which the microprocessor circuitry can operate at the desired clock frequency (while dissipating minimum energy) is accomplished by the circuit hardware.

Using closed loop feedback circuitry as shown in Figure 3.4, the supply voltage of a microprocessor can be dynamically adjusted [68]–[71]. The desired clock frequency ($f_{DESIRED}$) to execute a specific task is passed to a register by the OS. A replica of the most critical path in a microprocessor is employed to track the instantaneous clock frequency of a microprocessor at a specific supply voltage. A ring oscillator translates the supply voltage generated by an on-chip DVS DC–DC converter to a specific clock frequency (f_{CLOCK}). This clock frequency is compared to the desired clock frequency, generating a digital frequency error signal (f_{ERROR}). The loop filter using this error function generates the control signals for the drivers of the power transistors of the DC–DC converter to either modify or maintain the output voltage. The minimum supply voltage required for the operation of a microprocessor at a desired clock frequency is, thereby, dynamically generated.

The circuitry in a standard single supply voltage microprocessor system is typically designed to tolerate a maximum variation of approximately ±10% of the supply voltage. Alternatively, the circuitry of a DVS microprocessor is designed to operate over a much wider range of supply voltages to maximize the energy efficiency while maintaining the operation of the microprocessor system during supply voltage transients. Static CMOS logic gates

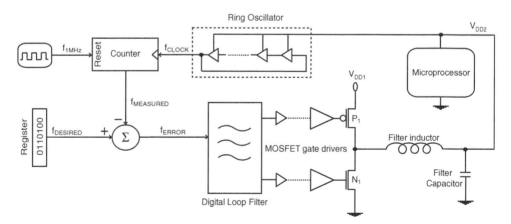

Figure 3.4   Feedback loop architecture for a dynamic voltage scaling circuit [70]

typically have a high tolerance to variations in the supply voltage. As the supply voltage is scaled, the delay of a static CMOS circuit scales proportionally [68], [70]. Dynamic CMOS circuits, custom arrays, latches, and analog circuits, however, cannot tolerate significant variations in the supply voltage. Modifications (such as replacing NMOS pass gates with transmission gates, employing wider keeper transistors in dynamic gates, and avoiding high stacks of transistors) to standard cell libraries are often necessary in the design of a DVS microprocessor [68]–[71].

The DVS circuit technique can also be used to effectively reduce die-to-die variations of the electrical characteristics such as the clock frequency and power consumption. In standard CMOS circuits, frequency binning is the primary method to enhance yield [98]. The frequency binning method reduces the clock frequency of those dies that do not satisfy the target active power requirements, thereby exploiting the linear dependence of dynamic switching power on the switching frequency. The reduction in frequency required to lower the active mode power below the maximum acceptable limit is typically significant. This significant reduction in clock frequency pushes the frequency distribution to the lower frequency bins, causing a violation of the minimum acceptable clock frequency for a large number of dies. The enhancement in yield with a standard frequency binning technique is, therefore, limited. Alternatively, the DVS circuit technique lowers both the supply voltage and clock frequency to reduce the active power. Since the dependence of the dynamic switching power on the supply voltage and frequency is cubic and the dependence of the leakage power on the supply voltage is superlinear, the DVS technique can satisfy target active power constraints with more moderate scaling of the clock frequency. The DVS technique, therefore, increases both the yield and the number of dies accepted in a higher frequency bin as compared to the standard frequency binning technique, subject to the constraints of total active power, burn-in leakage power, and standby leakage power consumption [98].

3.2 MULTIPLE SUPPLY VOLTAGE CMOS

Current ICs are typically designed to operate with a single supply voltage. The clock speed of a synchronous circuit is determined by the delay of the critical paths. A critical delay path between flip-flops FF_1 and FF_2 in a single supply voltage synchronous circuit is shown in Figure 3.5.

In a standard single supply voltage circuit, the value of the supply voltage is determined such that the target clock speed is achieved by the most critical (slowest) delay paths. However, as the number of critical paths typically constitutes only a small fraction of the total number of paths within an IC, a significant number of gates along the non-critical delay paths operate with excessive slack as shown in Figure 3.5 (signals propagate faster than necessary and arrive early, generating a time gap between the arrival and utilization of the input signals at the receiver end). If a signal arrives earlier at the receiver end of a non-critical path than is necessary, the performance of a circuit is not increased. Operating the gates along these non-critical delay paths at the same supply voltage level as the gates along the critical paths therefore wastes energy.

The multiple supply voltage circuit technique exploits these delay differences among the different signal propagation paths within an IC. The multiple supply voltage circuit technique selectively lowers the supply voltages of the gates on the non-critical delay paths while maintaining a higher supply voltage on the critical delay paths in order to satisfy a target clock

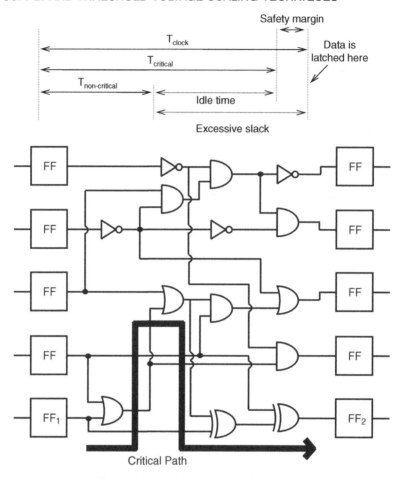

Figure 3.5   A single supply voltage circuit

frequency [28]. A dual supply voltage circuit in which the supply voltage of all of the gates along the non-critical delay paths are replaced by a lower supply voltage is shown in Figure 3.6.

Scaling the supply voltage of all of the gates along a non-critical delay path, however, may not always be feasible due to local timing constraints. The slack of a delay path after supply voltage scaling must be significantly lower but still greater than zero so as to not degrade the overall performance or reliability (by creating a race condition). A combination of high and low supply voltage gates can exist along a delay path if the delay requirements are not satisfied by scaling the supply voltages of all of the gates along a path. As discussed in Section 2.4, when a circuit supplied by a low supply voltage drives a CMOS circuit supplied by a higher supply voltage, static DC current and non-full rail output voltage swing problems occur. Specialized voltage-level converter circuits are required to interface the circuits operating at different supply voltages in a multiple supply voltage circuit [28], [32]. The power and area overhead of these voltage interface circuits must be included in the supply voltage optimization process. Circuit blocks with a lower supply voltage should, therefore, be chosen such that

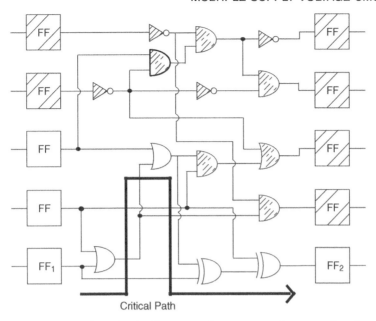

Figure 3.6   A dual supply voltage circuit. Those gates that operate at a lower supply voltage are shaded

the number of voltage interface circuits and the total power (including the power overhead of the voltage-level converters) are minimized while satisfying the timing constraints of all of the delay paths [28]. The clustered voltage scaling (CVS) technique, proposed in [28], minimizes the number of voltage-level converters in a multiple supply voltage circuit. In the CVS technique shown in Figure 3.7, the supply voltages are assigned such that no low supply voltage gate drives a high supply voltage gate.

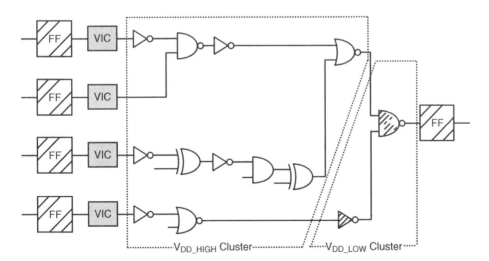

Figure 3.7   A dual supply voltage circuit with the clustered voltage scaling technique [28]. Those circuits operating at a lower supply voltage are shaded. VIC = Voltage Interface Circuit

A dual supply voltage media processor based on the CVS technique is presented in [28]. An automated synthesis method is proposed in [28] in which the total power dissipation is minimized without violating the timing constraints of the delay paths while scaling the supply voltage. A test circuit based on this method is fabricated in a 0.3 μm CMOS technology with a nominal supply voltage of 3.3 V. An interesting observation reported in [28] is that an optimum lower supply voltage exists which minimizes the total power in a dual supply voltage circuit. As the supply voltage of the gates along the non-critical delay paths is reduced, not only the dynamic switching energy consumption per circuit block but also the number of circuit blocks supplied by this lower supply voltage are reduced (due to the timing constraints). There is, therefore, an optimum low supply voltage that minimizes the total power (reported as 1.9 V in [28]). Within an optimum circuit configuration with the lowest power dissipation, the power supply for 76% of the cells is replaced with this lower supply voltage. A total of 5.8 million data paths are considered; fifteen thousand critical paths are identified, constituting only 0.3% of the total number of paths. As reported in [28], approximately 60% of the data paths have delays that are half of the cycle time (a 50% slack time) in a standard single supply voltage processor. A 39% to 57% reduction in power is reported with the dual supply voltage scheme as compared to a standard single supply voltage circuit operating at a nominal supply voltage of 3.3 V [28].

3.3 THRESHOLD VOLTAGE SCALING

Lowering the supply voltage is an effective way to reduce power consumption in CMOS ICs. Lowering the supply voltage, however, degrades circuit speed due to reduced transistor currents. The variation of the delay of an inverter with supply voltage for different threshold voltages, based on a 0.18 μm CMOS technology, is depicted in Figure 3.8.

Reducing the threshold voltages for a fixed supply voltage enhances circuit speed by increasing the gate overdrive ($|V_{GS}| - |V_t|$) of the transistors, as shown in Figures 3.8 and 3.9. Reducing the threshold voltage permits the supply voltage to be scaled without degrading speed. For example, as indicated with the delay line shown in Figure 3.8, the supply voltage can be scaled to 0.8 V from an initial voltage of 1.6 V and the threshold voltage can be scaled to 0.1 V from an initial voltage of 0.5 V while maintaining the same delay characteristics. If the ratio of the threshold voltage to supply voltage (V_t/V_{DD}) is maintained constant, the increase in delay due to scaling the supply voltage can be limited. The V_t/V_{DD} ratio should typically be maintained below 0.25 for reasonable performance in a scaled CMOS technology [67], [74], [82]. By scaling both the supply and threshold voltages, the power consumption and propagation delay of a CMOS circuit can be simultaneously reduced.

Although lowering the threshold voltage is effective in enhancing the speed, there are a number of issues that limit threshold voltage scaling in a new technology generation. Due to limitations in the maximum acceptable standby and active mode leakage power and die-to-die and within-die variations of the electrical characteristics, scaling the threshold voltage typically lags scaling the supply voltage in each new technology generation. Therefore, despite also scaling the threshold voltages, the V_t/V_{DD} ratio typically increases with technology scaling, degrading the achievable gain in circuit performance [74].

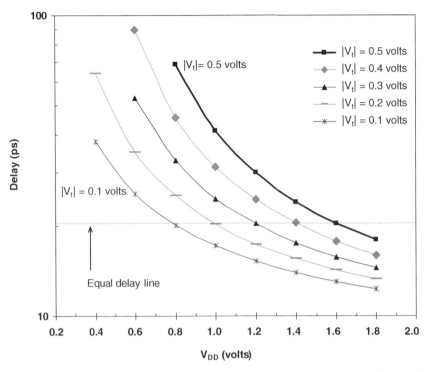

Figure 3.8   Variation of the delay of a CMOS inverter with supply voltage for different MOSFET threshold voltages (for a 0.18 μm CMOS technology)

A primary limitation to threshold voltage scaling is exponentially increasing subthreshold leakage currents with reduced threshold voltages. Subthreshold leakage current is the primary source of energy dissipation in an idle CMOS circuit. Energy consumption caused by standby leakage in portable devices is a significant concern since subthreshold leakage current can greatly reduce the lifetime of a battery. In a high performance IC, leakage currents in both the active and standby modes of operation are a serious concern, due to the aggressive scaling of the threshold voltages. Provided that current technology scaling trends continue, by 2010 more than half of the total active mode power consumption in high performance ICs is expected to be due to subthreshold leakage current, as shown in Figure 1.9 [5].

Another significant issue with threshold voltage scaling is the increasing effect of die-to-die and within-die parameter variations on the speed and power dissipation characteristics. Traditionally, the focus of semiconductor process development has been on controlling die-to-die parameter variations while within-die parameter variations have been somewhat neglected. Die-to-die parameter variations are caused by lot-to-lot and wafer-to-wafer differences in the processing temperature, wafer polishing, wafer placement, and the properties of the equipment used in the lithography process. Another source of die-to-die parameter variations is within-wafer differences primarily caused by aberrations in the stepper lens [64]. As the gate length of current semiconductor devices is lowered below the wavelength of light used in optical lithography (currently ranging from 193 nm to 248 nm), within-die parameter variations have also become a significant source of performance uncertainty in CMOS

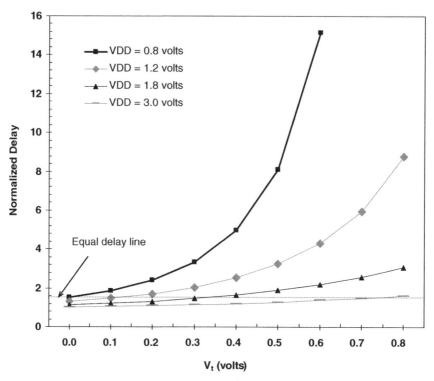

Figure 3.9   Effect of threshold voltage scaling on the delay of a 19 stage ring oscillator for four different supply voltages (for a 0.18 μm CMOS technology)

circuits. Die-to-die and within-die parameter variations such as variations in the critical dimensions (e.g., gate length, gate oxide thickness, and junction depletion width) are difficult to control and typically do not scale. Die-to-die and within-die fluctuations of the critical dimensions, therefore, effectively increase with technology scaling [63], [72].

Alternatively, as discussed in Section 2.2.1.1, the sensitivity of the threshold voltage to variations in the critical dimensions is greater due to increasing short-channel effects as the gate length is reduced with technology scaling. The doping concentration in the channel area is typically reduced to lower the threshold voltage of a MOSFET in a scaled CMOS technology. As shown in Figure 3.10, reducing the doping concentration in the channel area further increases short-channel and drain-induced barrier lowering effects. The sensitivity of the threshold voltage to variations in the critical dimensions, therefore, increases as the threshold voltage is scaled. Die-to-die and within-die fluctuations of the threshold voltages from a nominal target value increase with technology scaling.

Process parameter variations cause ICs to exhibit different clock frequency and power dissipation characteristics. The electrical characteristics of a CMOS circuit fabricated in a deep submicrometer process technology become increasingly non-deterministic due to the enhanced sensitivity of the devices to parameter variations. The variation in the performance characteristics increases with greater fluctuations in the threshold voltage. The number of dies that satisfies the minimum acceptable clock speed and the maximum tolerable power dissipation is reduced with scaled threshold voltages and minimum transistor dimensions,

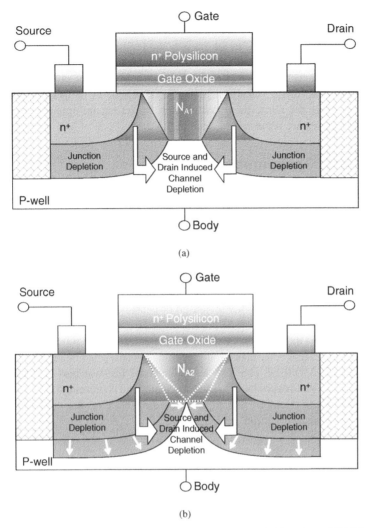

Figure 3.10 Effect of threshold voltage scaling on short-channel effects in an NMOS transistor. (a) A high-V_t short-channel MOSFET. (b) A low-V_t short-channel MOSFET. N_A = acceptor concentration in the channel area ($N_{A2} < N_{A1}$).

degrading the overall yield. The increasing cost of fabricating deep submicrometer ICs is, therefore, further aggravated by scaling the threshold voltages.

Several threshold voltage scaling techniques have been proposed that lower the effect of threshold voltage scaling on active and standby mode leakage power and die-to-die and within-die parameter variations in CMOS circuits. The body bias circuit technique dynamically changes the threshold voltage of the transistors by varying the voltage of the body terminal depending upon the dynamically changing power and speed requirements during circuit operation. The body bias circuit technique is discussed in Section 3.3.1. The multiple threshold voltage CMOS circuit technique employs transistors with different threshold voltages within the same circuit. A version of the multiple threshold voltage circuit technique reduces standby leakage power by employing high threshold voltage transistors between the power supply and ground terminals and the low threshold voltage circuitry. The high threshold voltage switches are cut off in the standby mode to suppress the high subthreshold leakage current characteristics of the low threshold voltage circuitry. An alternative dual threshold voltage circuit technique reduces the subthreshold leakage power by selectively employing low threshold voltage transistors along the critical delay paths and high threshold voltage transistors along the non-critical delay paths. Different multiple threshold voltage CMOS circuit techniques are reviewed in Section 3.3.2.

3.3.1 Body Bias Techniques

The exponentially increasing subthreshold leakage current and die-to-die and within-die threshold voltage variations determine the lowest acceptable (or achievable) threshold voltages for a specific deep submicrometer technology generation. The threshold voltage is typically adjusted during the fabrication process by varying the doping concentration in the channel area (see Figure 3.10). Alternatively, the body bias circuit technique utilizes the body terminal to dynamically modify the threshold voltage of a transistor during circuit operation. Depending upon the polarity of the voltage difference between the source and body terminals (V_{SB}), the threshold voltage can be either increased or decreased as compared to a zero body biased transistor.

The threshold voltage is increased when the source-to-substrate p–n junction of a MOSFET is reverse biased. The reverse body bias circuit technique is described in Section 3.3.1.1. The threshold voltage of a MOSFET can also be reduced by forward biasing the source-to-substrate p–n junction. The forward body bias circuit technique is presented in Section 3.3.1.2. The bidirectional body bias circuit technique (providing both reverse and forward body bias voltages) is described in Section 3.3.1.3.

3.3.1.1 Reverse Body Bias

The reverse body bias technique increases the threshold voltage of a MOSFET by applying a negative voltage across the source-to-substrate p–n junction, as shown in Figure 3.11. The

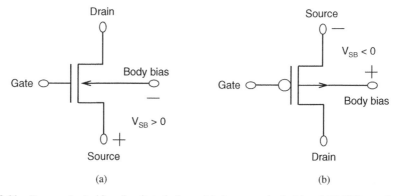

(a)   (b)

Figure 3.11   Reverse body bias circuit technique. (a) A reverse body biased NMOS transistor. (b) A reverse body biased PMOS transistor

variation of the charge distribution in the depletion region and inversion layer of a MOSFET under zero body bias and reverse body bias conditions is illustrated in Figure 3.12. In a MOSFET, the gate charge, insulator, mobile charges in the channel area, and the immobile ions in the depletion region form a capacitor (the MOS capacitor). The positive charge on the

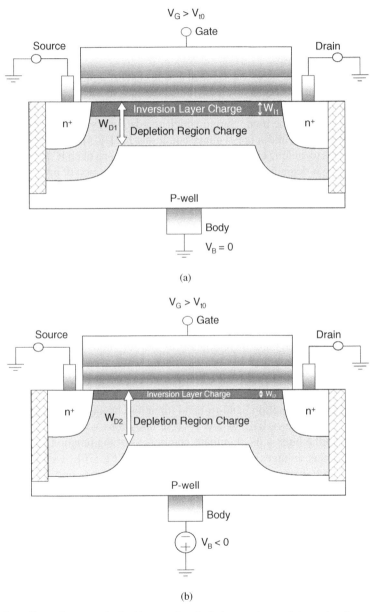

Figure 3.12 Effect of reverse body bias on the depletion region and inversion layer charge in a MOSFET. (a) A zero body biased NMOS transistor. (b) A reverse body biased NMOS transistor. $W_{D1} < W_{D2}$. $W_{I1} > W_{I2}$

gate is balanced by the sum of the electronic charge in the inversion layer and the negative ionic charge in the depletion region. When a MOSFET is reverse body biased, the width of the depletion region beneath the gate increases as shown in Figure 3.12(b). An increasing depletion width corresponds to an increase in the ionic charge in the semiconductor plate of the MOS capacitor. In order to maintain the charge balance, the mobile charge (number of electrons) in the inversion layer decreases, as depicted in Figure 3.12(b). As the number of mobile charges in the inversion layer is reduced in a reverse body biased MOSFET, the gate voltage needs to be increased to achieve a similar level of channel inversion as compared to a zero body biased MOSFET. The magnitude of the threshold voltage of a reverse body biased MOSFET, therefore, increases.

The reverse body bias technique can be used during standby and burn-in modes to increase the threshold voltage of selected or all of the transistors in an IC, thereby reducing the subthreshold leakage current. The standby mode is the mode during which a circuit is idle, while the burn-in mode is the mode during which standard stress tests are applied to an IC under elevated temperature and supply voltage conditions [72], [74], [75], [79]. Alternatively, the reverse body bias technique can be applied to the idle portions of an IC to reduce the active leakage power without degrading speed [76]. A significant reduction of up to 10 000 times in leakage power consumption is reported in [72] by applying a reverse body bias ($1.2V_{DD}$) to all of the transistors during the idle mode in a discrete cosine transform processor fabricated in a 0.3 μm CMOS technology.

Although increasing the reverse body bias voltage across the source-to-substrate p–n junction of a MOSFET increases the threshold voltage, thereby reducing the subthreshold leakage current, a reverse body bias also increases the tunneling leakage current at the reverse biased source-to-body and drain-to-body p–n junctions. The reverse biased junction leakage current in a MOSFET is composed of three primary components. The surface band-to-band tunneling current (also known as gate-induced drain leakage) is the dominant junction leakage current component at zero body bias and low junction temperature conditions. At high junction temperatures and zero body bias, the junction current due to the thermal emission of carriers is a significant leakage current component. When a reverse body bias is applied to a MOSFET at room temperature, junction leakage due to thermal emission is typically negligible as compared to the junction current due to band-to-band tunneling [75]. The junction band-to-band tunneling leakage current is dominated by gate-induced drain leakage (GIDL) at low reverse body bias voltages. The band-to-band tunneling current in the bulk is the dominant component of the junction leakage current at high reverse body bias voltages (typically above 0.5 V) [75], [76].

As the reverse body bias voltage is increased, both the surface and bulk band-to-band tunneling current components increase while the subthreshold leakage current decreases [74]–[76]. There is, therefore, an optimum reverse body bias voltage (specific to a process technology) that minimizes the total leakage power consumption [75]. The variation of the total standby power consumption of a test circuit for various body bias voltages is shown in Figure 3.13 [75]. The competing subthreshold and band-to-band tunneling leakage current mechanisms at increasing reverse body bias voltages are also illustrated. As shown in Figure 3.13, the total leakage power does not monotonically decrease with increasing reverse body bias voltage, due to increasing band-to-band tunneling current.

The reverse body bias technique has also been shown to be effective in reducing variations in the speed and power characteristics of ICs due to fluctuations in the supply voltage, temperature, and die-to-die process parameters [72], [73], [79], [80]. To compensate for the

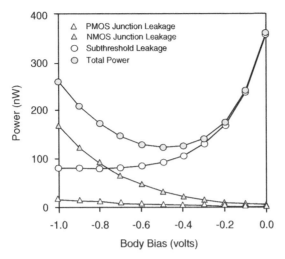

Figure 3.13 Variation of the total standby power of a microprocessor test circuit as a function of reverse body bias voltage [75]

unpredictable variations of these circuit parameters, an adaptive body bias control scheme, as shown in Figure 3.14, can be used. This adaptive reverse body bias circuit dynamically varies the body bias voltages depending upon local speed and power requirements. A feedback circuit integrated onto the same die as the IC tracks changes in speed and power consumption caused by variations in temperature, supply voltage, and/or process parameters. By matching an external reference signal to the delay or power information provided by a replica of a critical path, the necessary body bias voltages can be dynamically generated to adaptively compensate for variations in the circuit parameters. The external reference signal can be either a speed reference such as a clock signal (as shown in Figure 3.14) [73] or a power reference such as a target leakage current [72].

This speed-adaptive reverse body bias technique reduces die-to-die delay variations from 45% to 30%, as shown in Figure 3.15 [73]. Since die-to-die frequency variations are reduced, the adaptive reverse body bias technique provides an opportunity to further scale the threshold voltages without dissipating excessive leakage power. The worst case clock frequency

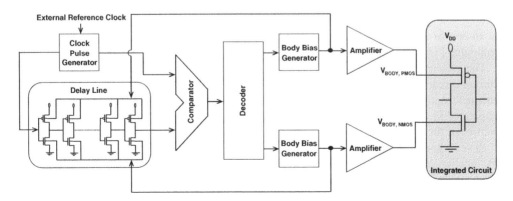

Figure 3.14 Block diagram of a speed-adaptive body bias circuit [79]

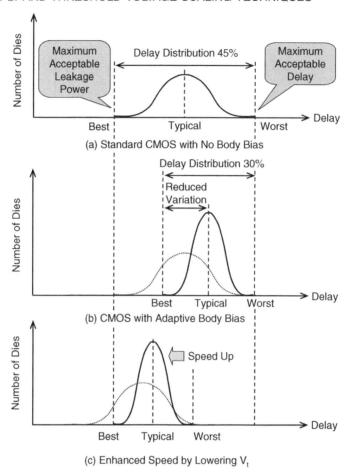

Figure 3.15   Reduced die-to-die delay variations by applying the speed-adaptive reverse body bias circuit technique to test circuits fabricated in a 0.25 μm CMOS technology [73]. (a) Delay distribution of standard CMOS circuits with zero body bias. (b) Reduced delay distribution with adaptive body bias. (c) Enhanced worst case speed by further scaling the threshold voltages with the adaptive body bias circuit technique

is shown to increase by up to 37% by applying this speed-adaptive reverse body bias technique [73].

The effectiveness of the reverse body bias technique to lower the subthreshold leakage current is reduced due to a weaker body effect with technology scaling [63], [74]–[78]. As the channel length is reduced, the body effect degrades due to increasing short-channel effects, as illustrated in Figure 3.16. Not only the gate terminal but also the body terminal lose some control of the charge distribution in the channel area in short-channel MOSFETs [63].

Reverse body biasing a MOSFET alleviates short-channel effects by increasing the width of the junction depletion regions. Moreover, in a circuit that is reverse body biased in the standby mode, the zero body bias threshold voltages are typically designed to be low to enhance the speed of the circuit when operating in the active mode. Lowering the doping concentration in the channel area to reduce the zero body bias threshold voltage further degrades the body

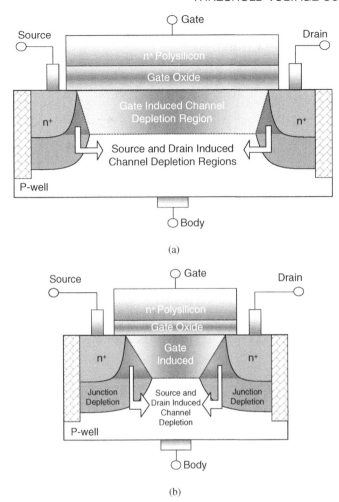

Figure 3.16 Body effect degradation due to channel length scaling. (a) A long-channel MOSFET. (b) A short-channel MOSFET

effect (see Figure 3.10) [63], [74]–[76]. The reverse body bias technique, therefore, becomes less effective in controlling the threshold voltage at reduced channel lengths and threshold voltages. Moreover, the optimum reverse body bias voltage that minimizes leakage current decreases due to increased band-to-band tunneling current with technology scaling [74]. Alternatively, the reverse body bias voltage necessary to achieve a target threshold voltage variation (ΔV_t) increases with technology scaling due to a reduced body effect. Hence, the higher the subthreshold leakage current becomes with technology scaling, the less effective the reverse body bias technique is in lowering leakage current.

   Another significant disadvantage of the reverse body bias technique is that variations in the leakage current due to parameter variations increase with higher reverse body bias voltages [76]. The threshold voltage becomes more sensitive to parameter variations due to increasing short-channel and drain-induced barrier lowering effects as the reverse body bias voltage is increased. The effect of the reverse body bias on short-channel effects

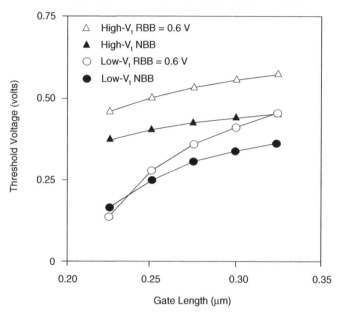

Figure 3.17   Increasing short-channel effects and threshold voltage roll-off with reverse body bias for low-V_t and high-V_t MOSFETs for a 0.25 μm CMOS technology [63]. NBB = No Body Bias, RBB = Reverse Body Bias

and threshold voltage roll-off is shown in Figure 3.17 [63]. As illustrated in Figure 3.17, low threshold voltage devices are more sensitive to variations in the critical dimensions. Threshold voltage roll-off further increases with higher reverse body bias voltages. Enhanced drain-induced barrier lowering with the reverse body bias circuit technique is illustrated in Figure 3.18.

3.3.1.2 Forward Body Bias

An alternative body bias scheme is the forward body bias technique. The threshold voltage of a MOSFET can be reduced by applying a positive voltage across the source-to-substrate p–n junction, as shown in Figure 3.19. The variation of the charge distribution in the depletion region and inversion layer of a forward body biased MOSFET as compared to a zero body biased MOSFET is illustrated in Figure 3.20. When a MOSFET is forward body biased, the width of the depletion region beneath the gate decreases, as shown in Figure 3.20(b). Reducing the depletion width corresponds to a decrease in the ionic charge on the semi-conductor plate of the MOS capacitor. In order to maintain charge balance, the mobile charge (number of electrons) in the inversion layer increases, as depicted in Figure 3.20(b). As the number of mobile charges in the inversion layer is increased, the gate voltage required to achieve a similar level of channel inversion as compared to a zero body biased MOSFET is reduced. The threshold voltage of a forward body biased MOSFET, hence, decreases.

As discussed in Section 3.3.1.1, for the purpose of reducing standby leakage current, the reverse body bias technique employs low threshold voltage transistors (under zero body bias conditions) to achieve a target circuit performance during the active mode of operation. The

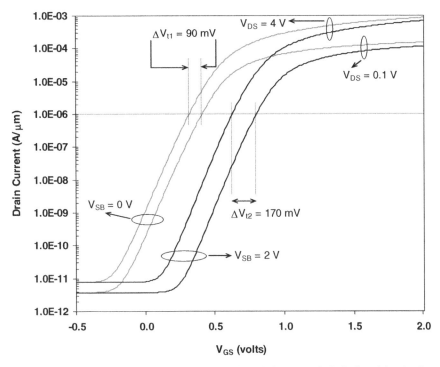

Figure 3.18 Effect of the reverse body bias circuit technique on drain-induced barrier-lowering ($\Delta V_t/\Delta V_{DS}$) for a 0.18 μm CMOS technology. The threshold voltage (V_t) is the gate-to-source voltage at which the drain current is equal to 1 μA/μm

threshold voltage of these transistors is increased during the standby and burn-in modes by applying a reverse body bias, thereby reducing the subthreshold leakage current. Alternatively, the forward body bias technique employs high threshold voltage ransistors (under zero body bias conditions) to maintain the standby leakage current below a target limit. The threshold voltage of these transistors is reduced during the active mode by applying a forward

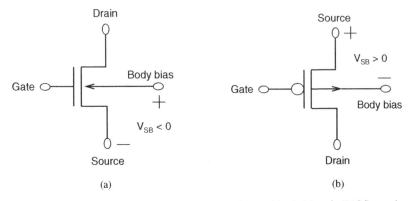

Figure 3.19 Forward body bias circuit technique. (a) A forward body biased NMOS transistor. (b) A forward body biased PMOS transistor

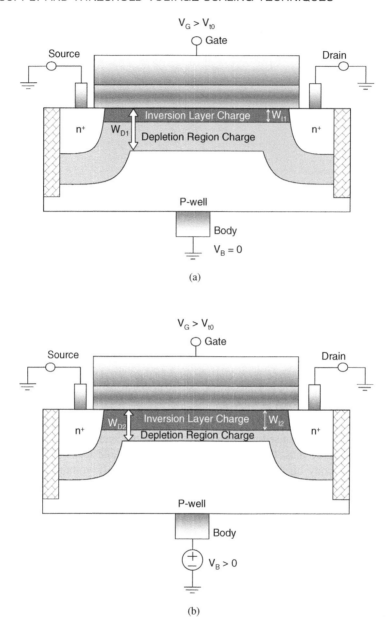

Figure 3.20 Effect of forward body bias on the depletion region and inversion layer charge in a MOSFET. (a) A zero body biased NMOS transistor. (b) A forward body biased NMOS transistor. $W_{D1} > W_{D2}$, $W_{I1} < W_{I2}$

body bias to achieve a target circuit speed. The forward body bias is removed (by applying either a zero body bias or a reverse body bias) during the standby and burn-in modes to increase the threshold voltages, thereby reducing the subthreshold leakage current [74], [77], [78].

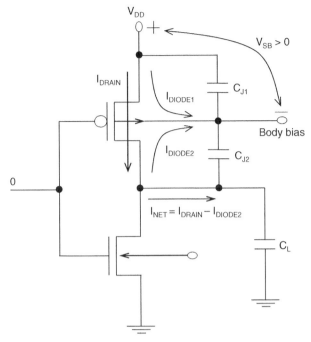

Figure 3.21 Schematic representation of a forward body biased CMOS circuit. I_{DIODE1} = source-to-body junction diode current, I_{DIODE2} = drain-to-body junction diode current, C_{J1} = source-to-body junction capacitance, and C_{J2} = drain-to-body junction capacitance

The maximum forward body bias voltage applicable to a MOSFET is limited by diode currents in the forward biased source-to-body and drain-to-body p–n junctions. The junction diode currents increase the active leakage power in a forward body biased circuit. The voltage swing at an output node can be degraded due to these junction diode currents if the forward body bias voltage is increased to effectively turn on the body diodes [78]. Moreover, as shown in Figure 3.21, the diode currents oppose the transition of the voltage state of a node, degrading the effective switching current and therefore the switching speed.

Another side effect of the forward body bias technique is the increased source-to-body and drain-to-body junction capacitances (C_{J1} and C_{J2} in Figure 3.21) with higher forward body bias voltages. A p–n$^+$ junction representing the body-to-source and body-to-drain junctions in an NMOS transistor is illustrated in Figure 3.22. Each diffusion area forms two p–n$^+$ junctions with the surrounding substrate (or well) and three p$^+$–n$^+$ junctions with the heavily doped channel stop implant areas in a MOSFET, as illustrated in Figure 3.23 [166].

The depletion width of a p–n junction is

$$x_d = \sqrt{\frac{2\varepsilon_{Si}}{q} \frac{N_A + N_D}{N_A N_D} (\phi_0 - V)}, \qquad (3.5)$$

$$\phi_0 = \frac{kT}{q} \ln\left(\frac{N_A N_D}{ni^2}\right), \qquad (3.6)$$

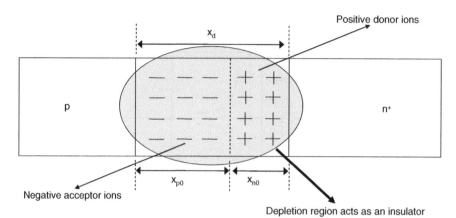

Figure 3.22 A p–n$^+$ junction. X_d = width of depletion region, x_{p0} = width of transition region into the p-side, and x_{n0} = width of transition region into the n-side

where ε_{Si} is the dielectric constant of silicon, k is the Boltzmann constant (1.38×10^{-23} J/K), T is the absolute temperature (K), q is the unit charge (1.6×10^{-19} C), N_A is the acceptor doping concentration in the p-side of the junction, N_D is the donor doping concentration in the n-side of the junction, n_i is the intrinsic carrier concentration of silicon, ϕ_0 is the built-in junction potential, and V is the voltage across the p–n junction. V is negative for reverse bias and positive for forward bias.

The depletion capacitance of an abrupt p–n junction is

$$C_j(V) = \frac{\varepsilon_{Si}A}{x_d} = A\sqrt{\frac{\varepsilon_{Si}q}{2}\frac{N_A N_D}{N_A + N_D}}\frac{1}{\sqrt{\phi_0 - V}}, \qquad (3.7)$$

where C_j is the p–n junction capacitance and A is the junction area. A reverse body bias increases the depletion width in the body diodes, thereby lowering the junction capacitances. Alternatively, since V is positive for a forward biased p–n junction, a forward body bias increases the junction capacitance by reducing the depletion width of the source-to-body and drain-to-body p–n junctions, as given by (3.5) and (3.7). These larger junction capacitances

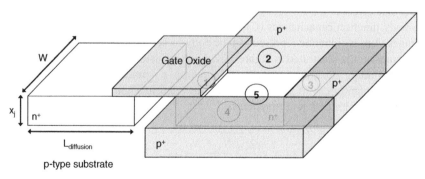

Figure 3.23 Body-to-drain (or body-to-source) p–n$^+$ junctions 1 and 5 with the p-type substrate and p$^+$–n$^+$ junctions 2, 3, and 4 with the p$^+$ channel stop implants in an NMOS transistor. W = transistor width, $L_{diffusion}$ = diffusion length, and x_j = diffusion depth

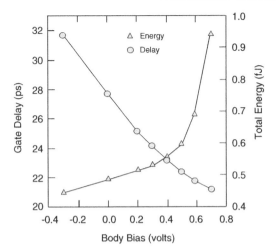

Figure 3.24 Variation of the propagation delay and energy consumption of a 101 stage ring oscillator with body bias voltage based on a 0.18 μm CMOS technology [77]

increase the active mode switching power and can become significant at high forward body bias voltages, increasing the propagation delay.

The variation of the propagation delay and active mode energy with the body bias voltage, for a 101 stage ring oscillator fabricated in a 0.18 μm CMOS technology, is illustrated in Figure 3.24 [77]. The energy–delay product of this ring oscillator is shown in Figure 3.25. The diode current from this test circuit is 3 nA/μm for a forward body bias voltage of 0.6 V. This diode leakage current is significantly smaller than the on-state drain current ($I_{DSAT} \approx 0.1$ mA/μm). Due to the significant enhancement of the drain current by the forward body bias circuit technique (e.g., the drain current increases by 40% for a 0.6 V forward body bias), the propagation delay is reduced with increasing forward body bias voltage despite the increasing diode currents and junction capacitances.

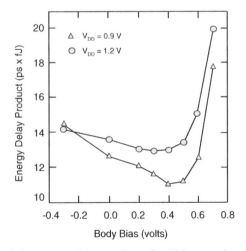

Figure 3.25 Variation of the energy–delay product of a 101 stage ring oscillator with body bias voltage based on a 0.18 μm CMOS technology [77]

As shown in Figure 3.24, the active mode energy increases approximately linearly up to a forward body bias voltage of 0.4 V due to the increasing junction capacitances. As the forward body bias voltage is increased beyond 0.4 V, the exponentially increasing junction diode currents significantly increase the active mode energy. Similarly, as shown in Figure 3.25, the energy–delay product is reduced with increasing forward body bias up to a bias voltage of 0.4 V. As the forward body bias voltage is increased beyond 0.4 V, the energy–delay product increases due to the significantly higher diode currents and junction capacitances. Similar optimum forward body bias voltages in the range of 0.4 V to 0.6 V have also been reported in [74] and [78] that maximize the clock frequency or minimize the energy–delay product.

An interesting observation reported in [74], [77], and [78] is that the speed enhancement and reduction in the energy–delay product achieved by the forward body bias technique increases with supply voltage scaling. As reported in [74], by applying a forward body bias voltage of 0.6 V, the oscillation frequency of a ring oscillator fabricated in a 100 nm CMOS technology is improved by 30% at a supply voltage of 1.5 V. Under the same body bias conditions, the speed enhancement increases to 45% and 150% as the supply voltage is scaled to 1.2 V and 0.8 V, respectively. The effectiveness of the forward body bias technique, therefore, increases provided that the supply voltage is scaled more aggressively than the threshold voltages, which is the likely trend in technology scaling (increasing the V_t/V_{DD} ratio due to the constraints imposed by higher standby leakage power and large manufacturing-induced variations in V_t). The junction capacitances and switching energy of this ring oscillator increase by 10% for a forward body bias voltage of 0.6 V (independent of the supply voltage) [74].

The forward body bias technique can also be used to reduce the active mode power consumption [74], [77], [78]. By forward biasing the substrate, a higher clock frequency can be achieved at a lower supply voltage. Lowering the supply voltage, due to the quadratic dependence of the switching energy on the supply voltage, significantly reduces the active mode power with moderate forward body bias voltages. As shown in [74], the active power consumption of a ring oscillator circuit (fabricated in a 100 nm CMOS technology) is reduced by approximately 40% by reducing the supply voltage from a nominal value of 1.1 V to 0.8 V. The same speed as a standard zero body biased circuit is maintained by applying a forward body bias voltage of 0.6 V. The power saving achieved by reducing the supply voltage outweighs the power overhead due to the increasing diode currents and junction capacitance.

Similarly, a microprocessor test circuit fabricated in a 0.15 μm CMOS technology operating at a nominal supply voltage of 1.2 V is reported in [78]. This test circuit operates at a clock frequency of 1 GHz under standard zero body bias conditions. It is shown that the supply voltage can be scaled to 1.1 V without degrading the clock frequency by applying a forward body bias of 0.5 V to all of the transistors within this IC. For this body bias condition, the active mode leakage current increases by approximately a hundred times. Similarly, the total switched capacitance increases by 10% [78]. The total active mode power dissipation (including the energy overhead due to the increasing active leakage current and junction capacitance) of this microprocessor is reduced by approximately 8%.

Similar to the reverse body bias technique, the effectiveness of the forward body bias technique is reduced with technology scaling due to the degradation of the body effect with increased short-channel effects at smaller channel lengths [74], [79]. However, unlike a reverse body biased transistor, the short-channel effects of a forward body biased transistor are lower as compared to a zero body biased transistor, as illustrated in Figure 3.26. As given by (3.5), when a forward body bias is applied to a MOSFET, the depletion width of the source-to-substrate and drain-to-substrate p–n junctions is reduced. Increasing the forward

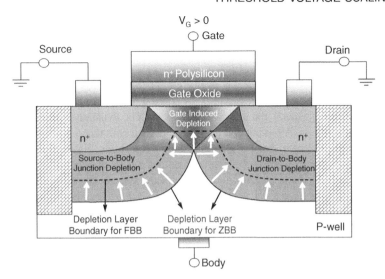

Figure 3.26 Effect of forward body bias on short-channel effects in an NMOS transistor. FBB = Forward Body Bias ($V_{Body} > 0$). ZBB = Zero Body Bias ($V_{Body} = 0$)

body bias, therefore, reduces short-channel and drain-induced barrier lowering effects while enhancing the body effect [74], [77], [78]. Moreover, contrary to the reverse body bias circuit technique, the zero body bias threshold voltages of a forward body biased circuit are typically higher, further enhancing the body effect while lowering the short-channel and drain-induced barrier lowering effects (due to the higher doping concentration in the channel area of a high threshold voltage transistor, as shown in Figure 3.10) [74], [77], [78]. The forward body bias technique, therefore, is more effective as compared to the reverse body bias technique with technology scaling. The forward body bias technique is expected to become more common as compared to the reverse body bias technique in future nanometer CMOS technology generations [77]–[79].

3.3.1.3 Bidirectional Body Bias

As described in [74], before the 0.13 μm technology node, a single static threshold voltage (no intentional body bias) has been standard for satisfying both speed and standby power requirements. However, due to the reduced circuit speed with lower supply voltages and the increased subthreshold and gate oxide leakage currents at scaled technologies, a single threshold voltage design space that satisfies both the speed and power requirements (for a single supply voltage system) is unlikely after the 0.13 μm technology generation. To scale the threshold voltages together with the supply voltages, some form of body bias is necessary. Either a reverse body bias or a forward body bias circuit technique will, therefore, be required to maintain the speed enhancements within a reasonable power budget below the 0.13 μm technology generation [74]. As the effectiveness of the reverse body bias circuit technique diminishes with technology scaling, a reverse-body-bias-only circuit technique will not simultaneously satisfy the speed and power requirements beyond the 70 nm technology generation [74]. Similarly, due to the weaker body effect, the forward-body-bias-only solution will no longer satisfy these performance requirements beyond the 50 nm technology generation. Beginning with the 50 nm technology generation, therefore, application of both forward and reverse body bias techniques within the same IC will become necessary to enhance circuit speed within a limited power budget [74].

In a bidirectional (forward and reverse) body bias circuit, the zero body bias threshold voltage of the transistors can be set to an intermediate value by controlling the channel doping concentration. In order to increase circuit speed, the threshold voltages can be dynamically reduced by forward body biasing the transistors. Alternatively, in order to reduce both the circuit speed and leakage power, the threshold voltages can be increased by reverse body biasing the transistors. As discussed previously, the forward body bias voltage that can be applied to a CMOS circuit is limited due to increasing diode currents and junction capacitances. Similarly, the reverse body bias voltage that can be applied to a CMOS circuit is limited due to the increasing junction band-to-band tunneling currents with technology scaling. Since the transistors in a bidirectional body biased circuit would be initially set to an intermediate threshold voltage (rather than the low threshold voltages utilized in a reverse-body-bias-only circuit or the high threshold voltages utilized in a forward-body-bias-only circuit), the bidirectional body bias technique can produce a wider choice of dynamically adjusted threshold voltages.

As discussed in Section 3.3.1.1, the adaptive reverse body bias technique can be used to reduce die-to-die parameter variations. However, as shown in [63] and [80], the reverse body bias circuit technique increases within-die parameter variations due to increasing short-channel effects as compared to a zero body biased circuit. Since the forward body bias technique reduces short-channel effects, a bidirectional adaptive body bias technique can be used to reduce both die-to-die and within-die parameter variations.

The measured clock frequency and leakage power consumption of 62 microprocessor test circuits fabricated in a 0.15 μm CMOS technology are shown in Figure 3.27 [80]. Due to

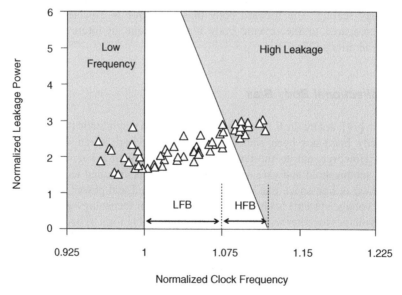

Figure 3.27   Leakage power and clock frequency characteristics of microprocessor test circuits fabricated in a 0.15 μm CMOS technology (LFB = Lower Frequency Bin, HFB = Higher Frequency Bin) [80]

die-to-die and within-die parameter variations, different dies display different frequency and leakage characteristics, as shown in Figure 3.27. For a manufactured IC to be marketable, both the minimum clock frequency and maximum leakage power requirements must be satisfied. The increasing die-to-die and within-die variations of the electrical characteristics are expected to further degrade yield with technology scaling. As shown in Figure 3.27, a significant number of dies are rejected due to violating either the speed or the power constraint.

A simple bidirectional adaptive body bias scheme that reduces die-to-die parameter variations is the application of a single adaptively generated body bias combination (for the NMOS and PMOS transistors) to an entire IC. This body bias combination can satisfy the delay requirements of the longest critical delay path in an IC using a feedback circuit similar to the circuit shown in Figure 3.14. Microprocessor test circuits based on this bidirectional adaptive body bias technique, fabricated in a 0.15 μm CMOS technology, are reported in [80]. With the standard zero body bias circuit technique, only 50% of the test circuits pass the speed and power tests. Moreover, most of the acceptable circuits have clock frequencies in the lower frequency bin (LFB). By applying a bidirectional adaptive body bias, the die acceptance rate increases to 100% [80].

This technique, however, ignores within-die parameter variations by applying a single body bias voltage combination to the entire IC. The die-to-die parameter variations similarly affect the electrical characteristics of all of the devices in an IC [64]. Applying a single set of body bias voltages to an entire circuit can, therefore, effectively reduce die-to-die parameter variations by shifting the threshold voltage of each of the devices by a similar ratio in the same direction. The within-die parameter variations, however, affect the electrical characteristics of all of the individual devices differently. Applying a single set of adaptive body bias voltages to an IC is, therefore, ineffective for reducing within-die parameter variations. Due to the significance of within-die parameter variations in a deeply scaled CMOS technology, despite the reduction of die-to-die variations and the yield enhancement to 100%, only 32% of the dies are acceptable in the higher frequency bin (HFB) by applying the same bidirectional adaptive body bias voltages to all of the transistors in an IC. An alternative adaptive body bias technique has been proposed in which a second set of test circuits are divided into different body bias zones. An independently generated adaptive body bias voltage is applied to each circuit zone, thereby lowering the within-die parameter variations. It has been shown that by applying this second bidirectional adaptive body bias technique, which reduces both the within-die and die-to-die parameter variations, the number of dies accepted in the highest frequency bin is increased to 99% while maintaining a 100% yield [80].

An alternative bidirectional body bias technique (named V_t-hopping) is proposed in [81] to lower the leakage power consumed during both the active and standby modes of operation. The V_t-hopping technique is essentially a dynamic threshold voltage scaling technique inspired by the DVS technique (DVS is discussed above in Section 3.1). The primary objective of the DVS circuit technique is to lower the dynamic switching power. The DVS circuit technique is effective in reducing the active power dissipation provided that the total active power is dominated by the dynamic switching power. Alternatively, the V_t-hopping scheme is effective in reducing the active power consumption provided that the dominant mechanism of active power consumption is subthreshold leakage current due to the aggressive scaling of the supply and threshold voltages ($V_{DD} \leq 0.5$ V and $V_t \approx 0$ V).

A reduced instruction set (RISC) microprocessor based on the V_t-hopping circuit technique fabricated in a 0.6 μm CMOS technology is reported in [81]. The V_t-hopping scheme utilizes two different sets of threshold voltages for operation at either a high clock frequency (f_{CLK}) or a low clock frequency ($f_{CLK}/2$). As the throughput requirement from a processor changes depending upon the variations of the workload, the operating system (OS) switches the desired clock frequency between f_{CLK} and $f_{CLK}/2$. The V_t-hopping circuitry, in response to a request from the OS, applies either a predetermined set of forward body bias voltages or a predetermined set of reverse body bias voltages to switch the operating frequency of the processor circuitry to f_{CLK} or $f_{CLK}/2$, respectively. The V_t-hopping technique reduces the total active power dissipation by up to 82% as compared to a low threshold voltage RISC processor for the same workload at a supply voltage of 0.5 V [81].

3.3.2 Multiple Threshold Voltage CMOS

Multiple threshold voltage CMOS technologies employ both high and low threshold voltage transistors within the same IC. The primary goal of multiple threshold voltage circuits is to selectively scale the threshold voltages together with the supply voltage in order to enhance circuit speed without significantly increasing the subthreshold leakage current.

The multiple threshold voltage circuit technique selectively places low threshold voltage transistors on the speed-critical paths of a circuit to enhance speed while operating at a reduced supply voltage [88]–[90]. The motivation for this multiple threshold voltage circuit technique is similar to the motivation for the multiple supply voltage circuit technique (discussed in Section 3.2). In a standard single threshold voltage circuit, the threshold voltages of the transistors are chosen to achieve a specific target clock frequency. Since the speed of a synchronous digital circuit is determined by the most critical (slowest) delay paths, the threshold voltages (similar to the supply voltage) are primarily chosen to lower the propagation delay of the signals along the critical paths in order to satisfy a target clock period. Since all of the transistors have the same nominal threshold voltage in a standard CMOS circuit, the signal propagation along many non-critical delay paths is unnecessarily fast, creating excessive slack. A single threshold voltage circuit essentially wastes power in the form of leakage current on many non-critical delay paths. The multiple threshold voltage circuit technique exploits this characteristic by selectively decreasing the threshold voltages only along the speed-critical paths. The primary objective of the multiple threshold voltage circuit technique is to minimize the number of low threshold voltage transistors required to satisfy a target clock frequency while maximizing the number of high threshold voltage transistors to achieve the lowest subthreshold leakage current. The multiple threshold voltage circuit technique provides an opportunity to further scale the threshold voltages (as compared to a standard single threshold voltage circuit) without violating any limitation in the total subthreshold leakage power. A target clock frequency can, therefore, be satisfied within a limited power budget by only scaling the threshold voltages of those portions of a circuit where a low threshold voltage transistor is required to achieve a specific propagation delay at a reduced supply voltage.

A dual threshold voltage PowerPC RISC microprocessor fabricated in a 0.25 μm dual threshold voltage CMOS technology is reported in [89]. Standard threshold voltage transistors are used in the caches, non-critical paths, and the leakage-sensitive dynamic circuits (primarily because of noise immunity concerns). The low threshold voltage transistors are selectively used on the speed-critical delay paths. The clock frequency of this dual threshold voltage microprocessor, with 40% of the transistors having a low threshold voltage, is enhanced by up to 10% as compared to a standard single threshold voltage microprocessor [89].

A 760 MHz G6 microprocessor fabricated in a dual threshold voltage 0.2 μm CMOS technology is reported in [90]. Similar to the previous example, standard threshold voltage transistors are used in the caches and the dynamic circuitry. The low threshold voltage transistors are selectively placed along the critical delay paths of the logic circuitry. The clock frequency of this microprocessor, with only 3% of the total transistors having a low threshold voltage (corresponding to 10% of the logic transistors), is enhanced by 10% as compared to a standard single threshold voltage microprocessor [90].

Another CMOS circuit technique employing multiple threshold voltage transistors, multi-threshold voltage CMOS (MTCMOS), has been proposed in [29]. An MTCMOS circuit is shown in Figure 3.28. In an MTCMOS circuit, all of the logic transistors have low threshold voltages to enhance circuit speed. In order to suppress the high subthreshold leakage current characteristics of the scaled low threshold voltage transistors, high threshold voltage switches are added between the low threshold voltage logic circuits and the power supply and ground lines. These high threshold voltage power supply and ground switches are controlled by a sleep signal. During the active mode of operation, the sleep control switches are activated, providing a virtual power and ground line for the logic circuits. During standby mode, these high threshold voltage sleep control switches are turned off, reducing the subthreshold leakage current. As shown in Figure 3.28, in a typical MTCMOS circuit, the same sleep control transistor is shared by several low threshold voltage logic gates, assuming

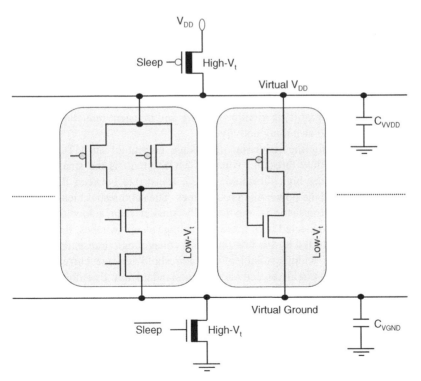

Figure 3.28 A multithreshold voltage CMOS (MTCMOS) circuit [29], [83]. The high threshold voltage transistors are illustrated by a bold line in the channel area

the switching activity of the gates connected to the same sleep switch is small (the switching activity is assumed to be less than 30% in [29]).

The delay of an MTCMOS circuit is degraded due to the high threshold voltage sleep switches connected in series, as compared to a standard low threshold voltage CMOS circuit. Due to the voltage drop across the series resistance of a sleep switch transistor, the voltage difference between the virtual power and ground lines is less than the standard full voltage swing between the primary power supply (V_{DD}) and ground. The effective supply voltage in an MTCMOS circuit is, therefore, smaller than the primary supply voltage V_{DD}. In order to minimize the speed degradation due to this reduction in the effective supply voltage, the width of the sleep switches is increased. Increasing the width of the sleep transistors, however, increases the area overhead, the subthreshold leakage current (which is proportional to the total width of the sleep switches), and the energy overhead of activating/deactivating the sleep switches. The optimum size of the high threshold voltage sleep switches is, therefore, a critical design issue in an MTCMOS circuit.

An average current method is proposed in [85] to optimize the width of the sleep transistors. This technique assumes that the current consumption of an MTCMOS circuit is constant and well understood before the beginning of the design process. An alternative technique is proposed in [86] to minimize the total width of the sleep switches. In this technique, the size of the sleep transistor of each low threshold voltage circuit is individually optimized based on circuit simulations. Once the optimum sleep transistor size is determined for all of the circuit blocks, the sleep transistors of the mutually exclusive gates (the gates that are guaranteed to not simultaneously switch) are merged, minimizing the total sleep transistor width [86]. Both of the circuit techniques described in [85] and [86] guarantee that for any input vector, the degradation in the effective supply voltage is limited such that the circuit delay is within a specific target range. For example, a maximum 2% degradation in effective supply voltage is allowed in [85] for a worst case 2% delay fluctuation. In addition to the resistive voltage drop, the virtual power and ground lines bounce whenever a gate switches. The parasitic capacitances of the virtual power and ground lines ($C_{Virtual-VDD}$ and $C_{Virtual-GND}$) temporarily supply charge to the internal logic circuits, limiting any transient reduction in the effective supply voltage due to the switching activity.

The primary reason for the reduction in leakage current of an MTCMOS circuit as compared to a standard low threshold voltage CMOS circuit is the smaller subthreshold leakage current due to the high threshold voltage transistors between the low threshold voltage logic circuits and the power and ground lines. The subthreshold leakage current of a high threshold voltage transistor is exponentially smaller than a low threshold voltage transistor. The serially connected high threshold voltage sleep switches, therefore, suppress the leakage current conducted by the low threshold voltage logic transistors. Moreover, the total effective transistor width available for subthreshold leakage current conduction is smaller in an MTCMOS circuit as compared to a standard low threshold voltage CMOS circuit, due to the sharing of the same sleep transistor by several logic gates. The total width of the sleep control switches is significantly smaller (typically less than 10%) than the total equivalent transistor width of the low threshold voltage logic gates. Reducing the total effective transistor width between the power supply and ground further decreases (linearly) the subthreshold leakage current.

The use of a narrower leakage current conduction path in order to reduce both the active and standby leakage power was first proposed by Sakata [84] in 1994. The switched power supply technique proposed in [84] assumes a single threshold voltage CMOS technology. This technique inserts a standard threshold voltage transistor, called a standby-current-limiting transistor, between the primary power supply and a virtual power supply line. A switched power supply circuit is divided into separate blocks to reduce the active leakage power in the unused sections of a circuit. Each circuit block has a separate block-select transistor between the virtual power supply line and the logic circuitry within the block. This switched power supply technique reduces the leakage current during both the active and standby modes of operation. During the active mode, the standby current limiting switch is always on and the block-select transistors of the unused circuit blocks are selectively turned off, reducing the active mode leakage current. During the standby mode, the standby current limiter switch is turned off, isolating the virtual supply line from the main power supply, thereby reducing the standby leakage current of the entire circuit [84]. The reduction in subthreshold leakage current with this technique is limited as compared to the MTCMOS technique, since the power supply switches have the same threshold voltage as the transistors within the logic circuits.

3.4 MULTIPLE SUPPLY AND THRESHOLD VOLTAGE CMOS

As discussed in Section 3.2, a dual supply voltage circuit technique offers significant power savings as compared to a standard single supply voltage CMOS circuit by exploiting the slack in a significant number of non-critical delay paths. Scaling the supply voltage along the non-critical delay paths, however, is limited as long as a standard single threshold voltage is maintained in a dual supply voltage circuit. To achieve higher power gains by further scaling the supply voltage, the threshold voltages must also be scaled. In an IC where the power consumption is dominated by dynamic switching power, a dual supply and threshold voltage circuit technique can significantly increase the power savings as compared to a dual supply and single threshold voltage circuit [92].

A typical high performance CMOS IC, such as a microprocessor, consists of two types of circuits which depend upon the circuit activity. A small portion of the circuitry (typically less than 5%) has a very high activity factor, on the order of 100% (such as clock circuits). These high activity circuits dominate the total power consumption of a typical CMOS circuit. The majority of the gates in a typical IC have a much lower activity factor, typically between 1% and 10% [53]. Although these low activity factor circuits constitute the large majority of the circuitry (typically greater than 95%), these circuits typically consume a smaller portion of the total power.

In a high activity circuit, the total power is typically dominated by the dynamic switching power even at very low supply and threshold voltages. Supply and threshold voltage scaling can, therefore, be effectively applied to a high activity circuit. For a low activity circuit, the total power is only dominated by the dynamic switching power at high supply voltages. As the supply voltage is decreased, the greater subthreshold leakage power begins to dominate the total power consumption. A low activity circuit, therefore, has a limited design space for supply and threshold voltage scaling. A multiple supply and threshold voltage circuit technique can exploit the different scaling characteristics of circuit blocks with different activity factors, as illustrated in Figure 3.29. Low supply and threshold voltages can be used in the high activity factor circuitry while high supply and threshold voltages can be maintained in the low activity factor circuitry. By scaling the supply voltage in the high activity factor circuitry which dominates the total power consumption of an IC, the total power can be significantly reduced (as compared to a single power supply system) without degrading the speed.

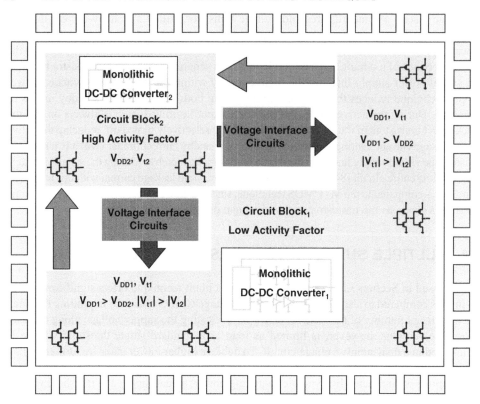

Figure 3.29   A multiple supply and threshold voltage IC with voltage partitioning based on the different activity factors among different circuit blocks

The power consumption of a test circuit with a varying supply voltage for three different activity factors assuming a 2 GHz clock frequency and a 100 nm CMOS technology is shown in Figure 3.30. As illustrated in Figure 3.30, for a 100% activity factor, the total power monotonically decreases as the supply and threshold voltages are scaled. For an activity factor of 10%, the total power is only dominated by dynamic switching power for supply voltages higher than 0.8 V. As the supply voltage is scaled below 0.8 V, the increasing subthreshold leakage power starts to dominate the total power consumption. Similarly, for an activity factor of 1%, the dynamic switching power dominates the total power dissipation for supply voltages above 1.2 V. As the supply voltage is scaled below 1.2 V, the higher subthreshold leakage current increases the total power consumption despite the reduced dynamic switching power.

A multiple supply and threshold voltage circuit technique is proposed in [53]. This technique employs a low supply voltage ($V_{DD2} = 0.5$ V) and low threshold voltage transistors ($|V_{t2}| \approx 0$) in the high activity circuitry. Alternatively, a high supply voltage and high threshold voltage transistors are employed in the low activity circuitry. The change in the total power consumption of the microprocessor test circuits based on dual supply and threshold voltages or standard single supply and threshold voltage circuit techniques with varying supply voltage is shown in Figure 3.31. The primary power supply (V_{DD1}) and threshold voltages ($|V_{t1}|$) are scaled for three different target clock frequencies.

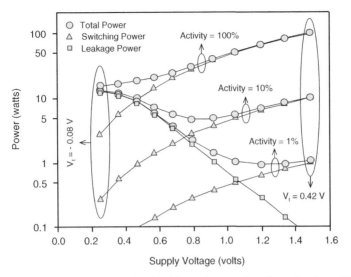

Figure 3.30 Power dissipation of a test circuit with varying supply voltage for three different activity factors assuming a 2 GHz clock frequency and a 100 nm CMOS technology [53]

As shown in Figure 3.31, for a target clock frequency of 2 GHz, the optimum supply voltage that minimizes the total power consumption of a dual supply and threshold voltage test circuit is 1.2 V. For this supply voltage and for the same clock frequency, the total power consumption of the dual supply and threshold voltage test circuit ($V_{DD1} = 1.2$ V,

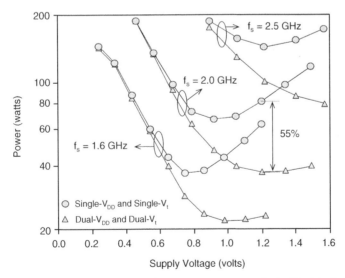

Figure 3.31 Power dissipation of dual V_{DD}/dual-V_t and standard single V_{DD}/single V_t test circuits with varying supply voltage for three different clock frequencies, assuming a 100 nm CMOS technology. For a dual V_{DD}/dual-V_t circuit, V_{DD2} and $|V_{t2}|$ are fixed at 0.5 V and 0 V, respectively. The supply voltage of the low activity circuits (V_{DD1}) is varied together with the threshold voltages ($|V_{t1}|$) while maintaining a target clock frequency [53]

$|V_{t1}| = 0.32$ V, $V_{DD2} = 0.5$ V, and $|V_{t2}| = 0$) is reduced by 55% as compared to a single supply and threshold voltage circuit ($V_{DD} = 1.2$ V and $|V_t| = 0.32$ V). The dual supply and threshold voltage circuit technique, therefore, enlarges the design space for supply voltage scaling, lowering the total power consumption as compared to a standard single supply and threshold voltage circuit while maintaining a target clock frequency.

3.5 DYNAMIC SUPPLY AND THRESHOLD VOLTAGE SCALING

As discussed in Section 3.1, DVS is an effective circuit technique for reducing the active mode power of an IC under varying workload conditions. It is assumed in the DVS circuit technique that the dominant power dissipation mechanism in a CMOS circuit is dynamic switching power. DVS adjusts the supply voltage of a circuit to the minimum voltage required to complete a specific task with a targeted latency, thereby exploiting the quadratic dependence of dynamic switching power on the supply voltage. The threshold voltages are preset during the fabrication process, typically to satisfy a clock frequency for anticipated maximum supply voltage and workload conditions. As the operating conditions (such as temperature) or workload changes, the supply voltage is varied to adjust the operating frequency while the threshold voltages are maintained the same. The DVS circuit technique, therefore, primarily focuses on minimizing dynamic switching power while ignoring the effect of subthreshold leakage current on the total power consumption of a CMOS circuit.

As the subthreshold leakage current is expected to become a significant contributor to the total power consumption of future nanometer CMOS circuits, dynamic switching and subthreshold leakage power must be balanced to minimize the total active power consumption [5], [97]. As the supply voltage is reduced, the threshold voltages must also be varied to maintain a target clock frequency while minimizing the total power consumption. The dynamic switching power is quadratically reduced and the subthreshold leakage power is exponentially increased with scaled supply and threshold voltages, respectively. As shown in Figure 3.32, the total power of a CMOS circuit does not monotonically decrease with smaller

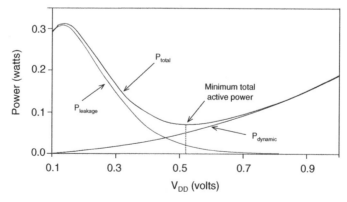

Figure 3.32 Active mode power dissipation of a CMOS circuit with varying supply voltage for a fixed operating frequency. The threshold voltages are modified together with the supply voltage to maintain a constant frequency [97]

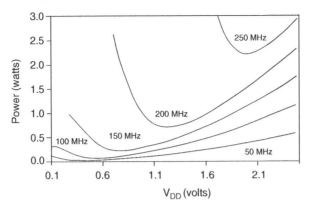

Figure 3.33 Active power with varying supply voltage for various clock frequencies. With each curve, the threshold voltages are modified together with the supply voltage to maintain a constant frequency [97]

supply voltages, provided that the circuit speed is maintained the same. There is an optimum supply and threshold voltage pair that minimizes the total active power of a CMOS circuit operating at a target clock frequency. As the required clock frequency changes with varying workload, the optimum choice of supply and threshold voltages that minimizes the total active power at a different target frequency also changes, as shown in Figure 3.33. Since the threshold voltages are optimized for a specific operating frequency and supply voltage in a DVS circuit, the DVS circuit technique is not capable of minimizing the active power consumption for clock frequencies other than a nominal (typically, the maximum) clock frequency.

The dynamic supply and threshold voltage scaling (DSTVS) circuit technique dynamically adjusts both the supply and threshold voltages to change the clock frequency whenever a change in the workload or operating conditions is detected [93], [97]. The DSTVS circuit technique offers flexibility to further reduce the total active mode power dissipation beyond the power savings achievable with the DVS technique by trading subthreshold leakage power for dynamic switching power.

Circuits operating at low or high clock frequencies require the threshold voltages to be significantly scaled as compared to nominal zero body bias threshold voltages. However, since the maximum body bias voltages that can be applied to a CMOS circuit are limited (see the discussion in Section 3.3.1), the range of threshold voltages that can be achieved by the body bias technique is also limited. The theoretical optimum supply and threshold voltages, therefore, may not always be achievable by applying a DSTVS technique [97].

3.6 CIRCUITS WITH MULTIPLE VOLTAGE AND CLOCK DOMAINS

Partitioning an IC into multiple circuit blocks can significantly enhance the effectiveness of dynamic voltage and frequency scaling techniques [155], [156]. Rather than determining an optimum supply voltage for an entire IC for a specific workload, the supply voltage and operating frequency can be optimized independently within each circuit block. This approach

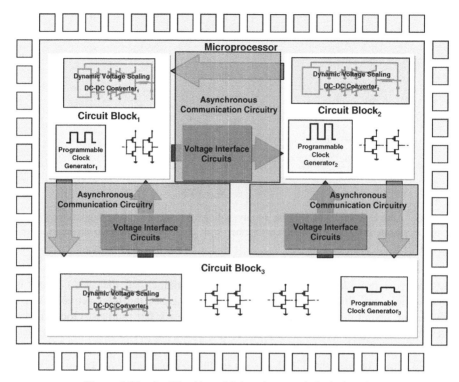

Figure 3.34   An IC with multiple voltage and clock domains

can be seen as a combination of globally asynchronous, locally synchronous (GALS) circuits and dynamic voltage and frequency scaling techniques. In addition to multiple clock domains that operate in a typical GALS-based circuit, different circuit blocks operating at different supply voltages exist in a multiple voltage and clock domain IC, as shown in Figure 3.34. In addition to lowering the power consumption, a multiple voltage and clock domain circuit can also significantly reduce the complexity of the on-chip synchronization circuitry.

A GALS microprocessor contains several independently clocked synchronous circuit blocks [174]. Within each circuit block (or clock domain), the clock distribution network is less complex due to the lower number of transistors and the smaller physical area. The tolerable clock skew and jitter within a clock domain can be achieved by adjusting the size and number of clock domains within a GALS-based IC. A GALS-based microprocessor is skew independent at the global level since communication among the clock domains does not require a global clock reference [174]. The clock domains within a GALS-based IC communicate via asynchronous handshaking protocols. In addition to reducing the power, jitter, and skew of the clock distribution network, GALS offers the opportunity to independently optimize the operating voltage and frequency within each clock domain. Since different clock domains communicate asynchronously, the clock frequency and supply voltage of each clock domain can be dynamically adjusted, satisfying the throughput needs of the system while minimizing the energy consumed by each circuit block.

3.7 SUMMARY

The effects of supply voltage scaling on the power consumption of a CMOS circuit are discussed in this chapter. Several circuit techniques that compensate for the degradation in speed at a lower supply voltage are also presented.

An effective strategy for lowering power consumption without degrading performance is to dynamically adjust the supply voltage as the workload varies with time. The objective of DVS is to provide high throughput when executing computation-intensive tasks while saving energy during other times by adequately lowering the supply voltage and operating speed of a circuit.

The multiple supply voltage CMOS circuit technique exploits the delay differences among the different signal propagation paths within an IC. This technique selectively lowers the supply voltage of the gates along the non-critical delay paths while maintaining a higher supply voltage along the critical delay paths in order to satisfy a target clock frequency.

The most widely employed technique for enhancing the speed of a circuit at a reduced supply voltage is to scale the threshold voltages. Lowering the threshold voltages enhances the gate overdrive ($|V_{GS}| - |V_t|$) of the transistors, thereby reducing the propagation delay of these circuits. Threshold voltage scaling, however, also increases the subthreshold leakage current, short-channel effects, and die-to-die and within-die parameter variations.

Several threshold voltage scaling techniques that lessen the deleterious side effects of threshold voltage scaling are discussed. The body bias circuit technique dynamically adjusts the threshold voltage of the transistors by varying the voltage of the body terminal depending upon the changing power and speed requirements during circuit operation. An alternative threshold voltage scaling technique, multiple threshold voltage CMOS, employs transistors with different threshold voltages within the same circuit. A version of the multiple threshold voltage circuit technique reduces the standby leakage power by employing high threshold voltage transistors between the power supply and ground terminals and the low threshold voltage circuitry. These high threshold voltage switches are cut off during standby mode to suppress the high subthreshold leakage current characteristics of the low threshold voltage circuitry. Another dual threshold voltage circuit technique reduces the subthreshold leakage power by selectively employing low threshold voltage transistors only along the critical delay paths and high threshold voltage transistors along the non-critical delay paths.

A promising circuit technique aimed at lessening the deleterious side effects of supply and threshold voltage scaling is the use of multiple supplies and threshold voltages. With this multiple supply and threshold voltage circuit technique, the supply and threshold voltages in the high activity circuitry are scaled to very low levels while high supply and threshold voltages are maintained in the low activity circuitry. The multiple supply and threshold voltage circuit technique significantly lowers the total power as compared to a standard single supply and threshold voltage circuit without degrading the clock frequency.

Since subthreshold leakage current is expected to become a significant contributor to the total power consumption of future nanometer CMOS circuits, dynamic switching and subthreshold leakage power must be managed in a coordinated way to minimize the total active power consumption. The DSTVS circuit technique dynamically adjusts both the supply and threshold voltages to change the clock frequency whenever a change in the workload or operating conditions is detected. The DSTVS circuit technique offers flexibility to further reduce the total active mode power consumption beyond the power savings achievable with the DVS technique by trading subthreshold leakage power for dynamic switching power.

Partitioning an IC into multiple supply voltage and clock domains can significantly enhance the effectiveness of dynamic voltage and frequency scaling techniques. Rather than determining an optimum supply voltage for an entire IC for a specific workload, the supply voltage and operating frequency can be optimized independently within each circuit domain. In addition to lowering the power consumption, a multiple voltage and clock domain circuit can also significantly reduce the complexity of the on-chip synchronization circuitry.

4 Low-Voltage Power Supplies

The speed and power characteristics of an IC are dependent on the supply voltage. For a circuit to operate reliably while satisfying target performance specifications, a stable supply voltage is essential. In most electronic systems (such as a computer system), several circuits with different voltage and current requirements exist. In order to supply these circuits with different voltages, currents, and power ratings, several voltage converters are necessary.

Many ICs (such as microprocessors, digital signal processors, dynamic random access memories, and static random access memories) are designed to operate with a DC supply voltage. Therefore, in a typical computer system, the AC voltage of the utility system is first converted into a DC voltage by an AC-to-DC converter. Once a DC voltage is obtained, several DC–DC converters generate the specific DC voltages required by the different circuit blocks within a system. The DC voltage supplied to a circuit must be maintained within a tight voltage envelope to satisfy the guaranteed performance and functionality of the circuitry under variations of the load current and DC input supply voltage. A power supply should, therefore, provide not only voltage conversion, but also voltage regulation. A DC–DC voltage regulator is a circuit that generates a regulated DC output voltage from a (possibly) unregulated DC input voltage with a different voltage magnitude and/or polarity.

The power from a voltage converter changes dramatically for different applications. In battery-operated portable applications, the power demand of a load is typically on the order of a few watts. Power supplies for computers and office equipment can supply hundreds to thousands of watts. In variable speed motor drives, the power required by a load ranges from kilowatts to megawatts. The power levels encountered in rectifiers and inverters that interface DC transmission lines to an AC utility system can be as high as thousands of megawatts [102]. The preferred circuit topology for voltage conversion in order to satisfy the voltage and current requirements of the load in an energy-efficient manner changes with the type of application. Many DC–DC conversion techniques have been developed over the years to provide energy-efficient voltage conversion for a wide variety of applications [102]–[115].

The systems of interest considered in this book are battery-operated portable devices and computer systems. In a computer system, various components operate with different voltage and current requirements. In order to ensure coherent operation of these circuits so that a stable system can be formed and maintained, high quality voltage regulation is necessary. The power supply system of a typical laptop computer is illustrated in Figure 4.1. A charger with a

Multi-Voltage CMOS Circuit Design V. Kursun and E. Friedman
© 2006 John Wiley & Sons, Ltd

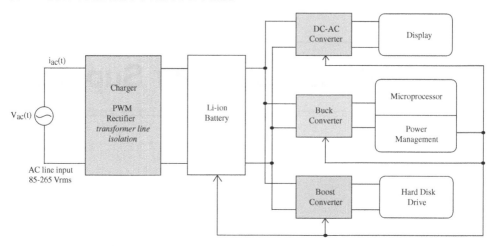

Figure 4.1   Power supply system for a laptop computer

transformer line isolation converts the AC line voltage to DC in order to charge the battery. A lithium-ion battery supplies unregulated voltage to the entire system. Several voltage converters generate the regulated supply voltages required by different circuit blocks from this unregulated battery input voltage. A buck converter produces the low DC voltage required by a microprocessor. A boost converter increases the battery voltage to the level required by the disk drive. A DC-to-AC converter produces the high frequency AC voltage that supplies the display [102].

A primary factor that determines the quality of a DC–DC converter is the output regulation, the stability of the output voltage over a wide range of input voltages and load currents. The output stability of a voltage regulator is characterized by the output voltage droop and peak-to-peak output voltage ripple under changing conditions of the load current and input voltage. Another important factor that determines the quality of a voltage converter is the energy efficiency of the voltage conversion process. A specific amount of energy is dissipated by the parasitic impedances of a DC–DC converter in order to generate a supply voltage. The choice of DC–DC conversion topology and related circuit techniques for a specific application is critical to the energy efficiency of the voltage conversion process. The energy efficiency η of a DC–DC converter is

$$\eta = \frac{P_{out}}{P_{in}} = \frac{V_{out}I_{out}}{V_{in}I_{in}}, \tag{4.1}$$

where P_{out} is the power supplied to the load, I_{out} is the load current, P_{in} is the total power supplied by the input power supply, and I_{in} is the current drawn from the input power supply. The power consumed by the parasitic impedances of the components within a voltage converter is

$$P_{lost} = P_{in} - P_{out} = P_{out}\left(\frac{1}{\eta} - 1\right). \tag{4.2}$$

Supply voltage scaling is an essential part of the technology scaling process in order to reduce the rate of increase of power consumption and to maintain device reliability with each new generation of high performance ICs. ICs, with higher power consumption and lower supply voltages, require innovative DC–DC conversion techniques that can provide substantial amounts of power and current with high energy efficiency [101]. DC–DC conversion techniques for low voltage ICs are examined in this chapter. The operating principles of linear, switched-capacitor, and switching DC–DC converters are described.

Linear regulators are used to generate a DC output voltage with a lower magnitude and the same polarity as compared to a DC input voltage. Linear regulators utilize resistive voltage division to produce an output supply voltage lower than an input supply voltage. Linear converters have intrinsically low efficiency, particularly if the input-to-output voltage conversion ratio is high [9], [102]. Linear regulators are found in many types of ICs due to the easy design, low circuit complexity, and small area consistent with an on-chip implementation [103]–[106]. The basic operating principles of linear regulators are presented in Section 4.1.

Switched-capacitor DC–DC converters (or charge pumps) are widely used in ICs to modify the amplitude and/or polarity of the primary power supply voltage of a system [9], [107]–[109]. Similar to a linear regulator, the efficiency of a switched-capacitor regulator is typically low. Alternatively, the area occupied by a switched-capacitor regulator is higher than a linear regulator. Unlike a linear regulator, a switched-capacitor DC–DC converter can change the polarity and increase the amplitude of an input supply voltage. Switched-capacitor regulators are, therefore, preferred in on-chip low-to-high voltage conversion or polarity reversing applications. The operating principles of switched-capacitor regulators are reviewed in Section 4.2.

Switching regulators are capable of modifying both the amplitude and polarity of the input voltages [9], [102], [112]–[115]. The primary advantages of a switching regulator are the high conversion efficiency and good output voltage regulation characteristics as compared to a linear or switched-capacitor DC–DC converter. The primary drawback of switching regulators, however, is the inductive elements (inductors and/or transformers) required for energy storage and filtering. Filter inductors are, to date, prohibitive in the fabrication of an on-chip switching DC–DC converter. The operating principles of switching regulators are discussed in Section 4.3. A summary of the low voltage DC–DC conversion circuit techniques presented in this chapter is provided in Section 4.4.

4.1 LINEAR DC–DC CONVERTERS

Linear (series-pass) DC–DC converters are popular due to the simple structure and small physical area of these circuits [9], [102]. Linear DC–DC converters operate on the principle of resistive voltage division. The operation of a simple linear voltage converter is illustrated in Figure 4.2.

As shown in Figure 4.2, in an ideal linear converter, the current supplied to the load is equal to the current drawn from the primary power supply V_{DD1}. The highest efficiency η_{max} attainable with an ideal (lossless) linear converter is, therefore,

$$\eta_{max} = \frac{V_{DD2}}{V_{DD1}}, \qquad (4.3)$$

where V_{DD2} is the DC output voltage supplied to the load and V_{DD1} is the DC input supply voltage. As given by (4.3), a linear DC–DC converter can only offer high energy efficiency (regardless of how ideal the circuit components are) if the difference between the input (V_{DD1}) and output (V_{DD2}) voltages is small.

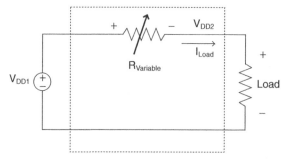

Figure 4.2   A simple voltage divider circuit describing the operating principle of a linear DC–DC converter

A linear voltage regulator should maintain the output voltage within certain (upper and lower) limits under variations of the load current and input supply voltage. A circuit schematic of a simple linear regulator with feedback circuitry for output voltage regulation is shown in Figure 4.3. A feedback circuit varies the gate voltage of a series transistor (which behaves as a variable resistor) by comparing the output voltage V_{DD2} to a reference voltage $V_{REFERENCE}$.

The ideal maximum efficiency, given by (4.3), is not attainable in a practical voltage regulator due to the energy losses in the parasitic impedances of the feedback circuitry and series switches. With careful design of the feedback circuit, however, the efficiency characteristics of a linear voltage regulator can approach the ideal upper limit. A different metric, called the current efficiency, is often used to characterize the efficiency of a linear regulator [106]. The current efficiency $\eta_{current}$ is

$$\eta_{current} = \frac{I_{out}}{I_{in}}, \tag{4.4}$$

where I_{out} is the current supplied to the load and I_{in} is the current drawn from the input power supply V_{DD1}. The current efficiency provides a measure of how close the efficiency of a linear regulator is to the ideal upper limit given by (4.3). The relationship between the energy and current efficiencies of a linear regulator is

$$\eta = \frac{V_{DD2}}{V_{DD1}} \eta_{current}. \tag{4.5}$$

As given by (4.5), if the difference of the output and input voltages is high, the energy efficiency of a linear regulator can be quite low despite a high current efficiency.

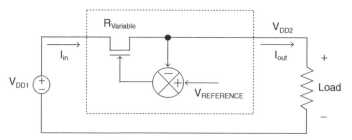

Figure 4.3   A linear voltage regulator

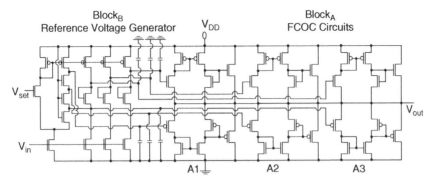

Block$_B$
Reference Voltage Generator   V$_{DD}$

Block$_A$
FCOC Circuits

V$_{set}$

V$_{in}$

V$_{out}$

A1   A2   A3

Figure 4.4   A high current–efficiency linear regulator [106]

A high current efficiency linear regulator is presented in [106]. This linear regulator circuit is shown in Figure 4.4. By a technique called "flexible control technique of output current (FCOC)" [106], the output current drive capability of a linear regulator is dynamically modified depending upon changes in the load current. The operation of the FCOC technique is illustrated in Figure 4.5.

The FCOC technique dynamically varies the output current in seven stages as determined by variations of the load current (see Figure 4.5). As shown in Figure 4.4, an FCOC linear regulator has three independent current mirror amplifiers (A1, A2, and A3) which are either turned on or turned off depending upon the relationship between the instantaneous output voltage and the six reference voltages generated by the reference voltage generator. The current mirror amplifier is shown in Figure 4.6. For each current mirror circuit, separate reference voltages ($V_{ref\text{-}high}$ and $V_{ref\text{-}low}$) are generated. When the output voltage is less than $V_{ref\text{-}low}$, N-Tr of the corresponding current mirror is turned on while P-Tr is turned off. When the output voltage is higher than $V_{ref\text{-}high}$, P-Tr of the corresponding current mirror is turned on while N-Tr is turned off. With this mechanism, the number of available current paths for charging or discharging the output node of the linear regulator is dynamically modified depending upon the variation of the load current. When the output voltage is within

Mode	Charging Mode				Discharging Mode		
Driving Current	3 x I	2 x I	I	= 0	I	2 x I	3 x I
Circuit Equivalent	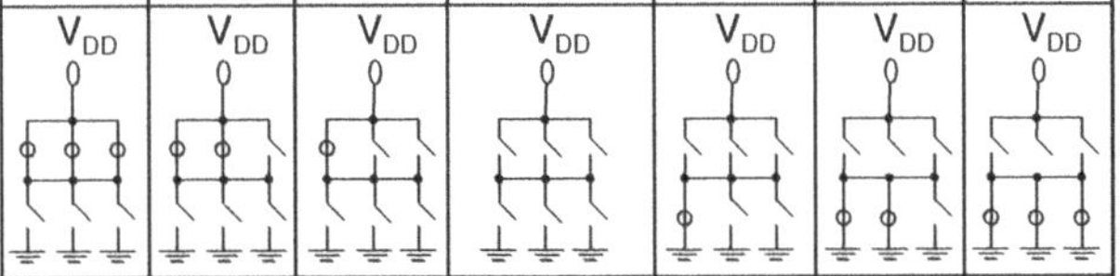 V$_{DD}$	V$_{DD}$	V$_{DD}$	V$_{DD}$	V$_{DD}$	V$_{DD}$	V$_{DD}$
V$_{out}$	$V_{ref\text{-}low\text{-}3}$ > V$_{out}$	$V_{ref\text{-}low\text{-}2}$ > V$_{out}$ > $V_{ref\text{-}low\text{-}3}$	$V_{ref\text{-}low\text{-}1}$ > V$_{out}$ > $V_{ref\text{-}low\text{-}2}$	$V_{ref\text{-}high\text{-}1}$ > V$_{out}$ > $V_{ref\text{-}low\text{-}1}$	$V_{ref\text{-}high\text{-}2}$ > V$_{out}$ > $V_{ref\text{-}high\text{-}1}$	$V_{ref\text{-}high\text{-}3}$ > V$_{out}$ > $V_{ref\text{-}high\text{-}2}$	V$_{out}$ > $V_{ref\text{-}high\text{-}3}$

Figure 4.5   Diagram representing the operation of the flexible control of the output current (FCOC) technique proposed in [106]

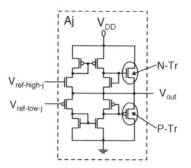

Figure 4.6 Current mirror amplifier

predetermined tolerable limits, all of the pull-up and pull-down transistors connected to the output node are cut off. This circuit technique, by dynamically varying the strength of a DC–DC converter to supply current based on varying load current requirements, reduces the power losses while stabilizing the output voltage over a wide range of load currents [106].

This linear regulator has been fabricated in a 1.2 μm CMOS technology [106]. For 5 V to 3 V conversion, the current efficiency is 96.5% (corresponding to an energy efficiency of 57.9%) at a DC output current of 5.7 mA. The current efficiency is reduced to 90% when the DC load current is increased to 27.35 mA. The fluctuation of the output voltage is between 2.85 V and 3.11 V as the input supply voltage changes from 4.5 V to 5.5 V. For load currents below 5.7 mA, the output voltage ripple (peak-to-peak) is less than 150 mV.

4.2 SWITCHED-CAPACITOR DC–DC CONVERTERS

Switched-capacitor DC–DC converters (or charge pumps) are used to generate a DC output supply voltage with a different magnitude and/or an opposite polarity as compared to a DC input supply voltage [9], [107]–[109]. On-chip switched-capacitor DC–DC converters are widely used to supply non-volatile memory circuits (flash and electrically erasable–programmable read only memories), dynamic random access memories (DRAMs), and analog portions of mixed-signal circuits [9], [107]. A schematic representation of a switched-capacitor DC–DC converter that doubles the input voltage is shown in Figure 4.7.

The operation of the switched-capacitor voltage converter circuit shown in Figure 4.7 behaves in the following manner. There are two mutually exclusive switching networks controlled by two phase control signals in a switched-capacitor DC–DC converter. Switches

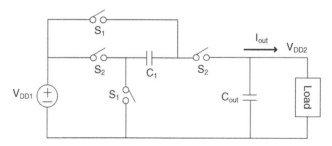

Figure 4.7 Schematic representation of a switched-capacitor DC–DC converter ($V_{DD2} = 2 \times V_{DD1}$)

labeled by one are controlled by the phase 1 control signal while the switches labeled by two are controlled by the phase 2 control signal. The phase 1 and phase 2 switch control signals do not overlap. When the phase 1 switches are activated (the phase 2 switches are cut off), C_1 is charged to V_{DD1}. In this phase, the output current is supplied by the output capacitor C_{out}. After C_1 is fully charged to V_{DD1}, the phase 1 switches are cut off and the phase 2 switches are activated. As a result of this connection, the output capacitor C_{out} is charged to twice ($2\times$) V_{DD1}. Provided that the switching action of the S_1 and S_2 switches is accomplished at a sufficiently high speed (the required frequency of switching depends upon the load current and output capacitor), the average output voltage (V_{DD2}) is maintained at twice V_{DD1}. An ideal step-up conversion ratio of two is, hence, achieved with the circuit illustrated in Figure 4.7. The output voltage in a practical switched-capacitor DC–DC converter based on the topology illustrated in Figure 4.7 is, however, less than $2 \times V_{DD}$ due to the voltage drop across the series resistance of the MOSFET switches. Moreover, in an actual charge pump, the output voltage degrades with increasing load current [107]–[109].

A primary disadvantage of a switched-capacitor DC–DC converter is the poor efficiency characteristics. The operation of a switched-capacitor regulator relies on periodically charging/discharging the charge pump capacitors through resistive switches. The internal power losses of a switched-capacitor regulator are, therefore, typically high [9], [107]–[109].

Another disadvantage of a charge pump circuit is the poor output regulation [9]. In order to maintain a steady DC output voltage, a certain amount of charge should be maintained across each charge pump capacitor. The only control mechanism that can be employed in a charge pump regulator to maintain a specific amount of charge in the charge pump capacitors under varying load current conditions is to vary the conductance of the switches charging/discharging the charge pump capacitors. This strategy, however, typically requires high energy consuming feedback circuitry, further degrading the efficiency of the switched-capacitor regulator. An energy-efficient feedback control scheme applicable to switched-capacitor regulators does not yet exist [9]. Switched-capacitor circuits are, therefore, typically used in applications with relaxed supply voltage constraints (such as DRAMs) that do not require tight voltage regulation [9], [107].

4.3 SWITCHING DC–DC CONVERTERS

A switching DC–DC converter generates a DC output supply voltage with a different magnitude and/or polarity than the DC input voltage. Among DC–DC converter topologies, switching voltage regulators are the most widely used due to the high efficiency and good output voltage regulation characteristics. Unlike a linear or switched-capacitor DC–DC converter, the efficiency of a switching DC–DC converter approaches 100% as the transistor switches are made more ideal (by employing a more advanced fabrication technology with reduced parasitic impedances).

Switching DC–DC converters can be divided into two primary categories. The first category of switching DC–DC converters utilizes transformers. Switching DC–DC converters with transformers are called isolated switching DC–DC converters [102]. The primary use of transformers in switching DC–DC converters is the DC isolation of the input and output grounds. Provided that the primary power supply operates at a relatively high voltage and/or is noisy, isolation of the load from the input supply is necessary to maintain reliable operation of the load. Another advantage of isolated switching DC–DC converters is the relatively easy and straightforward generation of multiple DC output voltages from a single DC input voltage. A single control circuit can be used to generate several different DC supply voltages by simply utilizing a multiple winding transformer, provided that the voltage regulation requirements of the load circuits are not excessively tight.

A second category of switching DC–DC converters utilizes inductors (no isolating transformers) for energy storage and signal filtering. These switching DC–DC converters without transformers are called non-isolated switching DC–DC converters [102]. Such converters are widely used in both low power and low voltage applications. A switching DC–DC converter that generates an output supply voltage with a higher magnitude as compared to the input supply voltage is a boost converter. Alternatively, a switching DC–DC converter that generates an output supply voltage with a smaller magnitude as compared to the input supply voltage is a buck converter. Buck and boost types of non-isolated switching DC–DC converters are widely used to generate voltage levels required by microprocessors, digital signal processors, memory modules, and hard disks in modern computer systems.

In a typical computer system, the power to a microprocessor is supplied by a buck converter. The operation of a buck converter is described in Section 4.3.1. Several power reduction techniques applicable to switching DC–DC converters are discussed in Section 4.3.2.

4.3.1 Operation of a Buck Converter

A buck converter is a standard switching DC–DC converter circuit topology with high efficiency and good output voltage regulation characteristics [9], [26], [30], [102], [112]–[115]. Buck converters are used to generate a DC output voltage from a higher DC input voltage with the same polarity. A buck converter is the preferred voltage regulator for a typical state-of-the-art high performance microprocessor. A buck converter circuit with a synchronous rectifier is shown in Figure 4.8. Traditionally, a Schottky diode is employed for rectification in a buck converter. However, as described in [110] and [111], in low voltage applications the overhead of the voltage drop across the diode p–n junction significantly degrades the efficiency. Therefore, in low voltage buck converters the diode is replaced by a MOSFET rectifier (N_1 in Figure 4.8) for enhanced efficiency [9], [102].

The operation of a buck converter circuit behaves in the following manner. The power MOSFETs, labeled as P_1 and N_1 in Figure 4.8, produce an AC signal at node$_1$ by a switching action controlled by a pulse width modulator (PWM). The AC signal at node$_1$ is applied to a second order low pass filter composed of an inductor and a capacitor. Assuming the filter corner frequency is much smaller than the switching frequency f_s of the power MOSFETs, the low pass filter passes to the output the DC component of the AC signal at node$_1$ and a small amount of high frequency harmonics generated by the switching action of the power MOSFETs.

The buck converter output voltage $V_{DD2}(t)$ is [26], [30]

$$V_{DD2}(t) = V_{DD2} + V_{ripple}(t), \tag{4.6}$$

where V_{DD2} is the DC component of the output voltage and $V_{ripple}(t)$ is the voltage ripple waveform observed at the output due to the non-ideal characteristics of the output filter. The

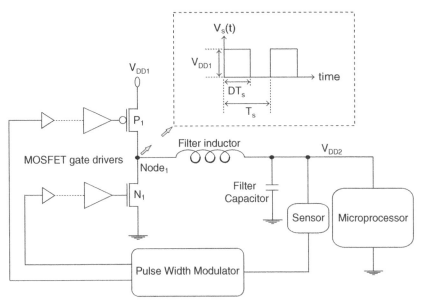

Figure 4.8   Buck converter circuit

DC component of the output voltage is [26], [30]

$$V_{DD2} = \frac{1}{T_s} \int_0^{T_s} V_s(t)dt = DV_{DD1},$$
(4.7)

where $V_s(t)$ is the AC signal generated at $node_1$. T_s, D, and V_{DD1} are the period, duty cycle, and amplitude, respectively, of $V_s(t)$. As given by (4.7), any positive DC output voltage less than V_{DD1} can be generated by a buck converter.

The power transistors are typically large in physical size and have high parasitic input capacitances. To control the operation of the power transistors, therefore, a series of MOSFET gate drivers is required. These gate driver buffers are typically tapered to drive these large capacitive loads [31], [120]. The gate driver buffers are controlled by a PWM. Using a feedback circuit, the PWM generates the necessary control signals for the power MOSFETs such that a square wave with an appropriate duty cycle is produced at $node_1$. During operation of a buck circuit, the duty cycle may be modified in order to maintain the output voltage at a desired value whenever variations in the load current and input voltage are detected. Due to the strong dependence of the output voltage on the switching duty cycle (see (4.7)), precise output voltage regulation can be produced by a buck converter with a fast feedback circuit [26], [30], [31].

The inductor current $i_L(t)$, output voltage $V_{DD2}(t)$, and capacitor current $i_C(t)$ waveforms of a buck converter are shown in Figure 4.9. The output voltage ripple is exaggerated in Figure 4.9 for better illustration. In a typical buck converter, the amplitude of the output voltage ripple ΔV_{DD2} must be maintained at a small level (typically less than 1%) as compared to the output DC voltage V_{DD1}.

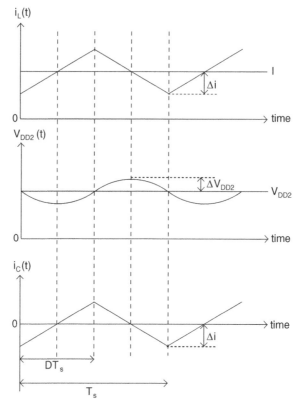

Figure 4.9 Inductor current $i_L(t)$, output voltage $V_{DD2}(t)$, and capacitor current $i_C(t)$ waveforms

The filter capacitance is chosen such that the impedance of the capacitor is much smaller than the load impedance. The AC component of the inductor current, therefore, passes through the filter capacitor while the DC component I passes through the load (see Figure 4.9). The output voltage increases whenever the inductor current rises above I, as the filter capacitor is being charged. Similarly, the output voltage falls whenever the inductor current decreases below I, as the filter capacitor is being discharged.

Expressions for the inductor current ripple Δi and the amplitude of the output voltage ripple ΔV_{DD2} (see Figure 4.9) are [26], [30], respectively,

$$\Delta i = \frac{(V_{DD1} - V_{DD2})D}{2L f_s},\tag{4.8}$$

$$\Delta V_{DD2} = \frac{(V_{DD1} - V_{DD2})D}{16LC f_s^2} = \frac{\Delta i}{8C f_s},\tag{4.9}$$

where L is the filter inductance, C is the filter capacitance, and f_s is the switching frequency.

4.3.2 Power Reduction Techniques for Switching DC–DC Converters

In low power portable systems the compactness and energy efficiency of a DC–DC converter is important due to the limitations of the available physical space, the limited effectiveness of the cooling solutions, and the need to extend battery life as much as possible. Switching DC–DC converters typically have large capacitive and inductive storage elements and power switches that occupy significant area and consume significant power. The size of the active and passive devices in a switching DC–DC converter are reduced with higher switching frequencies [26], [30], [31]. Increasing the switching frequency, however, also increases the power MOSFET-related switching losses [26], [30]. Increasing the switching frequency beyond a certain value, therefore, degrades the converter efficiency [26], [30].

Two techniques have been proposed for reducing the dynamic switching power of the power MOSFETs and are briefly discussed in this section. The zero voltage switching (soft switching) technique can substantially reduce switching losses associated with high frequency operation [9], [113]. In a zero voltage switching (ZVS) scheme, the filter inductor is used to charge/discharge the parasitic capacitances at the input of the output filter (node$_1$ in Figure 4.7) in a lossless manner. Provided that the activation times of the power transistors are carefully controlled, the parasitic capacitance at node$_1$ can be switched ideally without a power loss (neglecting the power dissipated by the series resistance of the filter inductor). If the power transistor P_1 (N_1) is turned on immediately after node$_1$ is charged (discharged) by the filter inductor, the power transistors are switched to a zero drain-to-source voltage difference, thereby eliminating the switching power losses that would have, otherwise, been dissipated in the power MOSFET while charging (discharging) node$_1$. The purpose of the power transistors in a ZVS voltage regulator is, therefore, to maintain the voltage of node$_1$ at either V_{DD1} or ground rather than charging or discharging node$_1$.

The activation time of the power transistors is critical for providing effective power savings with the ZVS circuit technique [9], [113]. The time required to charge/discharge node$_1$ depends upon the load current. To provide an effective ZVS over a wide range of loads, an adaptive dead time control scheme is proposed in [113]. The proposed technique dynamically adjusts the activation time of the power MOSFETs depending upon the instantaneous load current.

A similar power reduction technique, called the resonant gate drive technique, is proposed in [116] and [117] for lossless switching of the gate oxide related parasitic capacitance of the power MOSFETs. Based on a similar principle as the ZVS circuit technique, the resonant gate drive technique charges/discharges the input capacitance of the MOSFETs through an ideally lossless resonant circuit. Similar to the ZVS circuit technique, the resonant circuit technique stores energy in the inductors, utilizing this energy to charge or discharge the gate oxide-related parasitic capacitors. Provided that the activation time of the gate driver transistors are carefully controlled, the resonant gate drive technique can significantly reduce the power dissipated by the power MOSFET gate drivers [117].

4.4 SUMMARY

Three power supply topologies used in low voltage applications are reviewed in this chapter. The operating principles of linear, switched-capacitor, and switching DC–DC

Table 4.1 A Comparison of the Electrical Characteristics, and Typical Applications of Linear, Switched-Capacitor, and Switching DC–DC Converters

Type of DC–DC converter	Linear	Switched-capacitor	Switching
Low-to-high voltage conversion	No	Yes	Yes
High-to-low voltage conversion	Yes	Yes	Yes
Polarity reversal	No	Yes	Yes
Efficiency	Low	Low	High
Voltage regulation	Poor	Poor	Good
Area	Small	Medium	Large
Typical applications	DRAM	DRAM, flash, EEPROM, and mixed-signal	microprocessors, DSPs, SRAMs, and hard disks

converters are presented. A comparison of the electrical characteristics and typical applications of the linear, switched-capacitor, and switching DC–DC converters is listed in Table 4.1.

Linear regulators are used to generate a DC output voltage with a lower magnitude and the same polarity as compared to a DC input voltage. Linear regulators utilize resistive voltage division to produce an output supply voltage lower than an input supply voltage. Linear converters have intrinsically low efficiency, particularly if the input-to-output voltage conversion ratio is high. Linear regulators are found in many types of ICs (such as high density DRAMs) due to the easy design, low circuit complexity, and small area consistent with an on-chip implementation.

Switched-capacitor DC–DC converters (or charge pumps) are widely used in ICs to modify the amplitude and/or polarity of the primary power supply voltage of a system. Similar to a linear regulator, the efficiency of a switched-capacitor regulator is typically low. Alternatively, the area occupied by a switched-capacitor regulator is higher than a linear regulator. Unlike a linear regulator, a switched-capacitor DC–DC converter can change the polarity and increase the amplitude of an input supply voltage. Switched-capacitor regulators are, therefore, preferred in on-chip low-to-high voltage conversion or polarity reversing applications (such as flash and electrically erasable–programmable read only memories, DRAMs, and analog portions of mixed-signal circuits) where energy efficiency and tight output voltage regulation are not critical.

Switching regulators are capable of modifying both the amplitude and polarity of the input voltages. The primary advantages of a switching regulator are the high conversion efficiency and good output voltage regulation characteristics as compared to a linear or switched-capacitor DC–DC converter. A switching DC–DC converter is typically composed of discrete active and passive components and hence occupies a large area. The primary drawback of switching regulators is the inductive storage elements (inductors and/or transformers) required for energy storage and filtering. Filter inductors are, to date, prohibitive in the fabrication of an on-chip switching DC–DC converter.

A switching DC–DC converter that generates an output supply voltage with a higher magnitude as compared to the input supply voltage is a boost converter. Alternatively, a switching DC–DC converter that generates an output supply voltage with a smaller magnitude as compared to the input supply voltage is a buck converter. Buck and boost types of non-isolated switching DC–DC converters are widely used to generate voltage levels required by microprocessors, digital signal processors, memory modules, and hard disks in modern computer systems. In a typical computer system, the power to a microprocessor is supplied by a buck converter. The operation of a buck converter is described. Several power reduction techniques applicable to switching DC–DC converters are discussed.

5 Buck Converters for On-Chip Integration

Decreasing the power dissipation and current demand of high performance microprocessors are the two primary reasons for implementing a dual supply voltage (dual-V_{DD}) microprocessor [26], [30]. Due to the quadratic dependence of the dynamic switching power and the more than linear dependence of the subthreshold and gate oxide leakage power on the supply voltage, power dissipation is significantly reduced when portions of a microprocessor operate at a lower voltage level. A linear relationship exists between the current demand and power consumption of a microprocessor. Reducing the maximum power consumption, therefore, reduces the maximum current required by a microprocessor, thereby decreasing the number of power and ground pads on a microprocessor die. In order to maximize this reduction in current, the lower voltage supply of a dual-V_{DD} microprocessor should be integrated on the same die with the microprocessor. Moreover, in order to fully exploit expected reductions in power and current, the energy overhead of an integrated DC–DC converter to produce a second voltage level must be minimized.

Buck converters are popular due to the high efficiency and good output voltage regulation characteristics of these circuits [9], [26], [30]. In single power supply microprocessors, the primary power supply is typically an external (non-integrated) buck converter. In a dual-V_{DD} microprocessor, the choices are either a second external DC–DC converter, or a monolithic (both active and passive devices on the same die as the load) DC–DC converter.

In a typical non-integrated switching DC–DC converter, significant energy is dissipated by the parasitic impedances of the interconnect among the non-integrated devices (the filter inductor, filter capacitor, power transistors, and pulse width modulation circuitry) [9], [30]. Moreover, the integrated active devices of a pulse width modulation circuit are typically fabricated in an older technology with poor parasitic impedance characteristics [30].

Integrating a DC–DC converter with a microprocessor can potentially lower the parasitic losses as the interconnect between (and within) the DC–DC converter and the microprocessor is reduced [26], [30]. Additional energy savings can be realized by utilizing advanced deep submicrometer fabrication technologies with lower parasitic impedances. The efficiency

Multi-Voltage CMOS Circuit Design V. Kursun and E. Friedman
© 2006 John Wiley & Sons, Ltd

attainable with a monolithic DC–DC converter, therefore, is higher than a non-integrated DC–DC converter [26], [30].

Fabrication of a monolithic switching DC–DC converter, however, imposes a challenge as the on-chip integration of inductive and capacitive devices is required for energy storage and output signal filtering. Integrated capacitors and inductors above certain values are not acceptable due to the tight area constraints that exist within high performance microprocessor ICs. Another significant issue with integrated inductors is the poor parasitic impedance characteristics which can degrade the efficiency of a voltage regulator. The value, physical size, and parasitic impedances of the passive devices required to implement a buck converter, however, are reduced with increasing switching frequency [26], [30]. Integrated capacitors of small value (used for decoupling and constrained by the available area on the microprocessor die) are available in high performance microprocessors [118]. Furthermore, with the use of magnetic materials, a new integrated microinductor technology with relatively small parasitic impedances and higher cut-off frequencies (over 3 GHz) has recently been reported [119]. Therefore, employing switching frequencies higher than the typical switching frequency range found in conventional DC–DC converters permits the on-chip integration of active and passive devices of a buck converter onto the same die as a high performance microprocessor [30].

The efficiency characteristics of a buck converter, however, change dramatically as the switching frequency is increased. The switching frequency of DC–DC converters has been, so far, limited to the range from a few kilohertz to a few megahertz [9], [26], [30], [102], [112]–[115]. Based on oversimplified circuit models of switching DC–DC converters, a general assumption in the research community has been that a high switching frequency DC–DC converter is not feasible with the expectation that the efficiency would degrade significantly due to increased power losses at high switching frequencies [9], [102]. The low switching frequency range utilized in typical non-integrated DC–DC converters has been a result of this assumption rather than based on a study modeling the variation of the DC–DC converter efficiency as a function of the switching frequency. Comprehensive circuit models of the parasitic impedances of monolithic switching DC–DC converters are necessary in order to characterize an optimum circuit configuration with the maximum efficiency.

A parasitic model is presented in this chapter to analyze the frequency-dependent efficiency characteristics of a buck converter. A closed form expression that characterizes the power consumption of a monolithic buck converter is also described [26], [30]. The effects of scaling the active and passive devices and related switching and conduction losses on the total power characteristics of a buck converter are examined. With the presented buck converter energy model, a design space which characterizes the integration of both active and passive devices on the same die as a dual-V_{DD} microprocessor while maintaining high efficiency is determined for an 80 nm CMOS technology. An efficiency of 88.4% is shown for a voltage conversion from 1.2 V to 0.9 V while supplying 9.5 A maximum current. The area of the buck converter at the target design point is 12.6 mm^2 which is primarily occupied by a 100 nF filter capacitor. Full integration of a high efficiency buck converter on the same die as a dual-V_{DD} microprocessor is demonstrated to be feasible.

The parasitic circuit model and a closed form expression of the average power dissipation of a buck converter are presented in Section 5.1. With this analytic model, the efficiency characteristics of a buck converter are investigated in Section 5.2. Simulation results at a target design point are presented in Section 5.3. A summary of the research results presented in this chapter is offered in Section 5.4.

5.1 CIRCUIT MODEL OF A BUCK CONVERTER

A circuit model has been developed to analyze the frequency dependence of the efficiency characteristics of a buck converter. The circuit model for the parasitic impedances of a buck converter is shown in Figure 5.1.

The power consumption of a buck converter is a combination of the conduction losses caused by the parasitic resistive impedances and the switching losses due to the parasitic capacitive impedances of the circuit components. The power consumption of the pulse width modulation feedback circuit is typically small as compared to the power consumption of the power train (the power MOSFETs, MOSFET gate drivers, filter inductor, and filter capacitor) [26], [30], [31]. Only the power consumption of the power train components is, therefore, considered in the efficiency analysis.

MOSFET-related power losses are analyzed in Section 5.1.1. An analysis of the filter inductor-related losses is presented in Section 5.1.2. The filter capacitor-related losses are discussed in Section 5.1.3. An analytic expression for the total power consumption of a buck converter is presented in Section 5.1.4.

5.1.1 MOSFET-Related Power Losses

The total power loss of a MOSFET is a combination of conduction losses and dynamic switching losses. The conduction power is dissipated in the series resistance of the transistors operating in the active region. The dynamic power is dissipated each switching cycle while

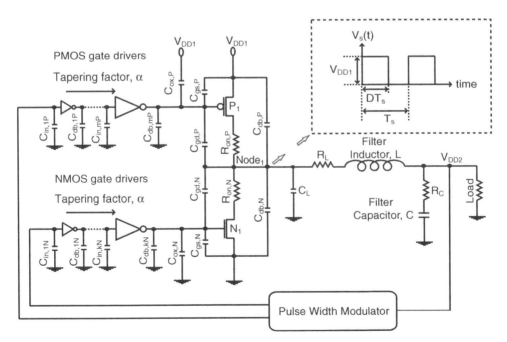

Figure 5.1 Circuit model of the parasitic impedances of a buck converter

charging/discharging the gate oxide, gate-to-source/drain overlap, and drain-to-body junction capacitances of the MOSFETs. In the following analysis it is assumed that the PWM control signals applied to P_1 and N_1 are non-overlapping. There is, therefore, no short-circuit current path through P_1 and N_1 during the PWM signal transition. The short-circuit power dissipated in the gate drivers is also neglected assuming the transition times of the input signal applied at each power MOSFET gate driver are smaller than the output transition times [9], [61], [120].

The average power consumption of a power MOSFET and the related gate drivers is

$$P_{MOS} = \frac{R_0}{W} i_{rms}^2 + EWf_s, \tag{5.1}$$

$$E \cong \frac{\alpha}{\alpha - 1}(C_{ox} + C_{gs} + 2C_{gd} + C_{db})V_{DD1}^2, \tag{5.2}$$

where P_{MOS} is the total power consumed during a switching cycle of a power MOSFET (which includes the power dissipated by the MOSFET gate drivers), R_0 is the equivalent series resistance of a 1 μm wide transistor, i_{rms} is the rms current passing through the power MOSFET, W is the width of the power MOSFET, α is the tapering factor of the power MOSFET gate drivers, C_{ox}, C_{gs}, C_{gd}, and C_{db} are the gate oxide, gate-to-source overlap, gate-to-drain overlap, and drain-to-body junction capacitances, respectively, of a 1 μm wide MOSFET, and E is the unit energy (per 1 μm wide power MOSFET) consumed during a full switching cycle of a power MOSFET (including the energy dissipated in the gate drivers).

As given by (5.1), increasing the MOSFET transistor width reduces the conduction losses while increasing the switching losses. An optimum MOSFET width, therefore, exists that minimizes the total MOSFET-related power. The optimum MOSFET width and power loss expressions for a target rms current and switching frequency are

$$W_{opt} = \sqrt{\frac{R_0 i_{rms}^2}{f_s E}}, \tag{5.3}$$

$$P_{MOS}(\text{min}) = 2\sqrt{R_0 i_{rms}^2 f_s E}. \tag{5.4}$$

As previously mentioned, it is assumed that the PWM signals for the power MOSFETs are non-overlapping. The time period during which both N_1 and P_1 are cut off is called the dead time. The rms currents through N_1 and P_1 (assuming a small dead time to switching period (T_s) ratio as compared to D) are

$$i_{rms}(NMOS) = \sqrt{(1 - D)\left(I^2 + \frac{\Delta i^2}{3}\right)}, \tag{5.5}$$

$$i_{rms}(PMOS) = \sqrt{D\left(I^2 + \frac{\Delta i^2}{3}\right)}, \tag{5.6}$$

where I is the DC current supplied to the load.

Applying (5.4) for N_1 and P_1 and substituting the rms current expressions (5.5) and (5.6), an expression for the total MOSFET-related optimized power consumption of a buck converter $P_{tot,MOS}(opt)$ is

$$P_{tot,MOS}(opt) = a\sqrt{\left(I^2 + \frac{\Delta i^2}{3}\right)f_s},$$

(5.7)

$$a = 2\left[\sqrt{R_{0\,NMOS}(1-D)E_{NMOS}} + \sqrt{R_{0\,PMOS}\,DE_{PMOS}}\right].$$

(5.8)

5.1.2 Filter Inductor-Related Power Losses

Some portion of the total energy consumption of a buck converter is due to the series resistance and the stray capacitance of the filter inductor. Integrated spiral inductors have a high series resistance and other intrinsic problems associated with a planar design which makes these inductors area and energy inefficient [119]. Integration of a spiral inductor with sufficient inductance is, therefore, not feasible for a high performance microprocessor. A novel low resistance inductor has previously been reported [119]. Assuming the inductor parasitic impedances scale linearly with the inductance [121], the total power dissipated in the filter inductor is

$$P_{tot,inductor} = b\left[\frac{I^2}{\Delta i f_s} + \frac{\Delta i}{3 f_s} + \frac{C_{L0}V_{DD1}^2}{R_{L0}\Delta i}\right],$$

(5.9)

$$b = \frac{(V_{DD1} - V_{DD2})DR_{L0}}{2},$$

(5.10)

where C_{L0} and R_{L0} are, respectively, the parasitic stray capacitance and parasitic series resistance per nH inductance.

5.1.3 Filter Capacitor-Related Power Losses

The filter capacitance affects the total power consumption of a buck converter due to the effective series resistance (esr) R_C. Assuming the integrated capacitor is implemented utilizing the gate oxide capacitance of a MOSFET, the total power dissipation of a filter capacitor is

$$P_{tot,capacitor} = d f_s \Delta i,$$

(5.11)

$$d = \frac{8R_{0cap}L_{cap}C_0\Delta V_{DD2}}{3},$$

(5.12)

where R_{0cap} is the effective series resistance of a 1 μm wide MOSFET, C_0 is the gate oxide capacitance per μm^2, and L_{cap} is the channel length of the MOSFET.

5.1.4 Total Power Consumption of a Buck Converter

Combining (5.7), (5.9), and (5.11), the total power consumption of a buck converter is

$$P_{buck} = a\sqrt{\left(I^2 + \frac{\Delta i^2}{3}\right)f_s} + b\left[\frac{I^2}{\Delta i f_s} + \frac{\Delta i}{3 f_s} + \frac{C_{L0}V_{DD1}^2}{R_{L0}\Delta i}\right] + d f_s \Delta i, \tag{5.13}$$

where a, b, and d are given by (5.8), (5.10), and (5.12), respectively.

The power dissipation of a buck converter is a strong function of the switching frequency and the inductor current ripple. As given by (5.13), a higher switching frequency increases the MOSFET and filter capacitor-related losses while decreasing the filter inductor-related losses. Similarly, the MOSFET and filter capacitor power losses increase with greater inductor current ripple. The relationship between the inductor losses and the inductor current ripple, however, is more complicated. Increased current ripple reduces the filter inductance required for a target switching frequency, which reduces the inductor parasitic impedances and the related power loss. A higher current ripple, however, also increases the rms current through the filter inductor which causes the conduction losses of the inductor to be larger.

Depending upon the ratio of the inductor- and MOSFET-related components of the total power consumption of a buck converter, the efficiency can actually increase with higher switching frequency and current ripple within a specified (f_s, Δi) range. This observation agrees with the analysis presented in Section 5.2.

5.2 EFFICIENCY ANALYSIS OF A BUCK CONVERTER

The efficiency of a buck converter η is

$$\eta = 100 \times \frac{P_{load}}{P_{load} + P_{buck}}, \tag{5.14}$$

where P_{load} is the average power delivered to the load and P_{buck} is the average total internal power consumption of a buck converter as given by (5.13).

The DC–DC converter efficiency is strongly dependent on the switching frequency f_s. The switching frequency is, therefore, a primary design variable in this analysis. High f_s is desirable for a monolithic buck converter due to the dependence of the filter inductance and capacitance on f_s as described by (4.8) and (4.9). As f_s is increased, values of L and C required to satisfy the target output voltage and current are reduced. Since the integration of the active and passive devices of a buck converter circuit is a primary concern in this analysis, a frequency range higher than the typical range found in conventional buck converters is used throughout the analysis. The range of switching frequency f_s is varied from 10 MHz to 4 GHz.

As given by (5.13), another buck converter circuit parameter that strongly affects the circuit efficiency is the inductor current ripple Δi. For a target f_s, increasing Δi reduces the required filter inductance (see (4.8)). The filter capacitance, however, must be increased to maintain the output voltage ripple ΔV_{DD2} within acceptable limits with increased Δi for a target f_s (see (4.9)). An appropriate Δi, therefore, should be chosen that results in a filter inductance and capacitance suitable for on-chip integration.

In the following analysis, it is assumed that the two power supply voltage levels used in the microprocessor are 1.2 V (V_{DD1}) and 0.9 V (V_{DD2}). The maximum load current demand I is assumed to be 9.5 A. It is also assumed that the tapering factor α of the power MOSFET drivers is two for a worst case energy efficiency analysis. It should be noted that the optimal tapering factor of the power MOSFET gate drivers for higher energy efficiency is typically much greater than the tapering factor assumed in this analysis (see Chapter 6). An 80 nm CMOS technology is assumed.

The global maximum efficiency circuit configuration is discussed in Section 5.2.1. The effect of a reduced filter capacitance on the circuit configuration and the resulting efficiency characteristics of a buck converter are analyzed in Section 5.2.2. The allowable output voltage ripple ΔV_{DD2} is assumed to be 5 mV in Sections 5.2.1 and 5.2.2. Another advantage of an integrated DC–DC converter is that a higher ΔV_{DD2} is acceptable as compared to a non-integrated DC–DC converter, while satisfying the same load voltage and current specifications. The beneficial effects of increasing ΔV_{DD2} on the efficiency characteristics of a buck converter are examined in Section 5.2.3.

5.2.1   Circuit Analysis for Global Maximum Efficiency

The power consumption and efficiency variation of a buck converter are shown in Figures 5.2 and 5.3, respectively, for $0.1\,\text{A} \le \Delta i \le 9.5\,\text{A}$ and $10\,\text{MHz} \le f_s \le 4\,\text{GHz}$. The 'z' axis represents the power (in watts) and the efficiency (%) in Figures 5.2 and 5.3, respectively. The MOSFET, filter inductor, and filter capacitor components of the total power dissipation of a buck converter are shown in Figures 5.4–5.6, respectively. The 'z' axis in Figures 5.4–5.6 represents the power (in watts).

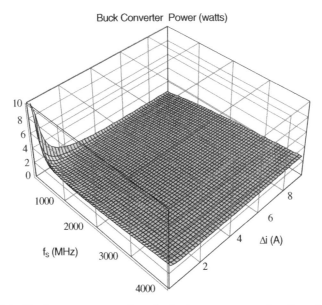

Figure 5.2   Total power consumption of a buck converter as a function of f_s and Δi

Buck Converter Efficiency (%)

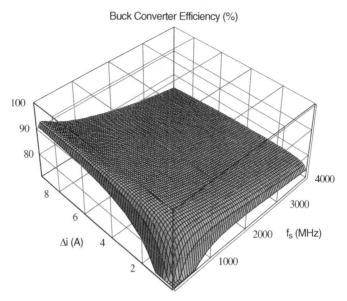

Figure 5.3   Efficiency of a buck converter as a function of f_s and Δi

As shown in Figures 5.4 and 5.6, the MOSFET- and capacitor-related power increases with increasing switching frequency and inductor current ripple. Alternatively, as shown in Figure 5.5, the inductor power monotonically decreases with increasing switching frequency and inductor current ripple. The capacitor power is negligibly small (less than 1%) as compared to the inductor and MOSFET power over the entire (f_s, Δi) range of analysis.

MOSFET Power (watts)

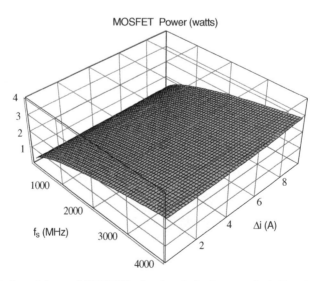

Figure 5.4   Variation of the total MOSFET-related optimized power (including the power dissipated in the gate driver buffers of the power MOSFETs) with the switching frequency and inductor current ripple

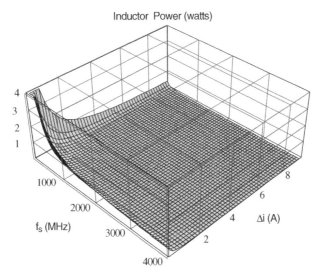

Figure 5.5   Variation of the total power dissipated in the filter inductor with the switching frequency and inductor current ripple

The filter capacitor losses, although included in the analysis, are therefore not further discussed in the chapter.

The efficiency of a buck converter is characterized by competing inductor and MOSFET losses. At low f_s and Δi, the buck converter power is primarily dissipated in the filter inductor. As the switching frequency and current ripple are increased, the inductance is dramatically reduced, lowering the parasitic losses of the inductor. The MOSFET power increases, however, with increasing f_s and Δi. At a certain range of f_s and Δi, the inductor losses dominate the total

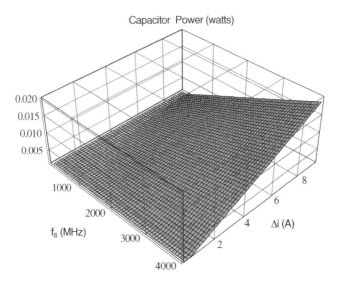

Figure 5.6   Variation of the total power dissipated in the filter capacitor with the switching frequency and inductor current ripple

losses. As shown in Figure 5.2, the total power dissipation of a buck converter decreases with increasing f_s and Δi in the range dominated by the inductor losses. After the peak efficiency is reached, increasing MOSFET losses begin to dominate the total power dissipation of a buck converter. Hence, the efficiency degrades with further increases in f_s and Δi.

An optimum switching frequency and inductor current ripple pair exists that maximizes the efficiency of a buck converter. The global maximum efficiency is 92% at a switching frequency of 114 MHz and a current ripple of 9.5 A. The required filter capacitance and inductance at this operating point are 2083 nF and 104 pH, respectively. This filter capacitor would occupy an unacceptably large area on a microprocessor die for the target technology. Fabrication of a monolithic DC–DC converter at this maximum efficiency operating point is, therefore, not feasible.

5.2.2 Circuit Analysis with Limited Filter Capacitance

Because of the area overhead of an integrated capacitor, the filter capacitance that can be integrated on a microprocessor die is limited. The filter capacitance is swept between 100 nF and 1 nF to evaluate the effects of a reduced filter capacitance on the circuit configuration and the efficiency characteristics of a buck converter. The circuit configurations at each operating point offering the highest efficiency (η) are listed in Table 5.1.

As listed in Table 5.1, an efficiency of 88.4% can be achieved with a 100 nF filter capacitance. The area occupied by the maximum efficiency configuration with a 100 nF filter capacitance is 12.6 mm^2. The maximum achievable efficiency is reduced to 74.7% as the filter capacitance is lowered to 1 nF. The reason for the increase in power dissipation with reduced filter capacitance is explained by the relationship between the filter inductor, filter capacitor, output voltage ripple, and the inductor current ripple, as described by (4.8) and (4.9). As the filter capacitance is reduced, the filter inductance and switching frequency are both increased to satisfy the output voltage and current requirements. Therefore, both the switching and conduction power dissipation of the power MOSFETs and the filter inductor increase with reduced filter capacitance, thereby degrading the converter efficiency. Note that the conduction and switching components of the MOSFET power dissipation are equal at the optimum transistor width. Both power components increase due to increasing f_s and MOSFET series resistance R_{on} as the filter capacitance is reduced.

With this analysis, a design space is presented that supports full integration of a high efficiency buck converter onto a microprocessor die. With further capacitor space available on the microprocessor die, the attainable efficiency increases toward the global maximum efficiency of 92%, as described in Section 5.2.1. Another advantage of a higher filter capacitance is the lower switching frequency requirement, thereby improving circuit reliability and making the design of the pulse width modulation circuitry less complicated.

Table 5.1 Maximum Efficiency Circuit Configurations of a Buck Converter with Different Filter Capacitances

C (nF)	η (%)	f_s (MHz)	L (pH)	W_{P1} (mm)	W_{N1} (mm)
1	74.7	3174	279	50.8	20.2
10	82.8	1227	187	81.7	32.5
100	88.4	477	124	131.9	52.5

5.2.3 Output Voltage Ripple Constraint

In an external (non-integrated) DC–DC converter, as the current demand of the micropro-cessor varies during operation with changing circuit activity level, the voltage supplied to the load also varies due to the resistance of the interconnect between the converter output and the microprocessor package, the microprocessor input pads, and the on-chip power distribution network. A droop window of 10% is, typically, allowed as the microprocessor current demand steps from a minimum (caused by standby leakage current) to a maximum. The external wiring (the interconnect between the converter output and the on-chip power distribution network) that exists in an external DC–DC converter does not occur in an on-die DC–DC converter. A larger portion of the acceptable 10% voltage drop window can therefore be applied to the output voltage ripple of an integrated DC–DC converter.

The effect of increasing the output voltage ripple on the circuit configuration and efficiency characteristics of a buck converter is examined in this section. The output voltage ripple ΔV_{DD2} is increased from 5 mV (the value assumed in Sections 5.2.1 and 5.2.2) to 25 mV. The filter capacitance C is also increased from 1 nF to 100 nF. The maximum efficiency attainable with each ΔV_{DD2} and C pair is shown in Figure 5.7(a). The switching frequency and filter inductance of the buck converter circuit configuration offering the highest efficiency are shown in Figures 5.7 and 5.8, respectively. The filter inductor and MOSFET components of the total power dissipation of a buck converter are illustrated in Figure 5.8(b).

As shown in Figures 5.7 and 5.8, increasing the output voltage ripple reduces the switching frequency and filter inductance required to satisfy the DC–DC converter output voltage and current specifications for a fixed filter capacitance. With decreased switching frequency and filter inductance, both the MOSFET- and inductor-related components of the total power dissipation of a buck converter are reduced, as shown in Figure 5.8(b). The efficiency attained by a limited filter capacitance, therefore, increases by relaxing the output voltage ripple constraint. Moreover, as the required filter inductance is reduced, the die area required for the integrated filter inductor becomes smaller. Similarly, as the required switching frequency is reduced, the circuit reliability increases while the design of the pulse width modulation circuit becomes less complicated. As shown in Figure 5.7(a), the maximum achievable efficiency increases by up to 7.9% as the output voltage ripple is increased from 5 mV to 25 mV. Similarly, the filter inductance and switching frequency required for a corresponding maximum efficiency configuration are reduced by 24% and 48.7%, respectively, as ΔV_{DD2} is increased from 5 mV to 25 mV.

5.3 SIMULATION RESULTS

The buck converter circuit configuration that produces the maximum efficiency (see Table 5.1) with a filter capacitance of 100 nF and an output voltage ripple of 5 mV is evaluated assuming an 80 nm CMOS technology. The analytic expression (see (5.13)) for the total power consumption of a buck converter is effective in estimating the circuit efficiency characteristics. The buck converter efficiency as determined by simulation at the target design point is 86%, only differing by 2.4% from the efficiency determined from the analytic expression.

The converter output voltage (which supplies 9.5 A of DC current to the load) is shown in Figure 5.9(a). The peak-to-peak output voltage ripple is actually lower than the analytic expectation of 10 mV. This behavior is noted since the voltage drop across the equivalent

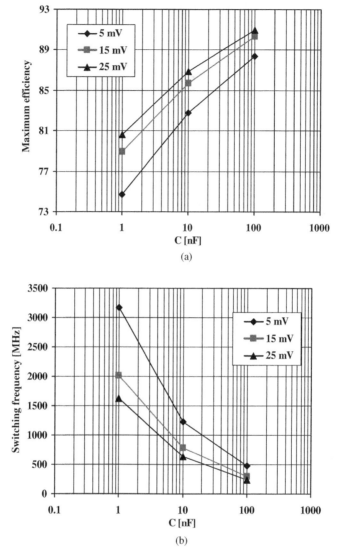

Figure 5.7 Variation of maximum efficiency and switching frequency of a buck converter with filter capacitance C ($1\,\text{nF} < C < 100\,\text{nF}$) and output voltage ripple ΔV_{DD2} ($5\,\text{mV} < \Delta V_{DD2} < 25\,\text{mV}$). (a) Maximum efficiency. (b) Switching frequency

parasitic resistance of the power MOSFETs and the filter inductor has been neglected during the steady-state analysis used in the development of (4.8) and (4.9).

The response of the buck converter to changes in the current demand (between the minimum and the maximum) at the load has also been evaluated. A 10% output voltage window is allowed as the average current demand of the microprocessor swings from a minimum (I_{min}) to a maximum (I). The minimum current demand I_{min} is caused by leakage current when the microprocessor is idle and is assumed to be 25% of the maximum current demand I [26], [30]. The waveforms illustrating the DC–DC converter output response for a

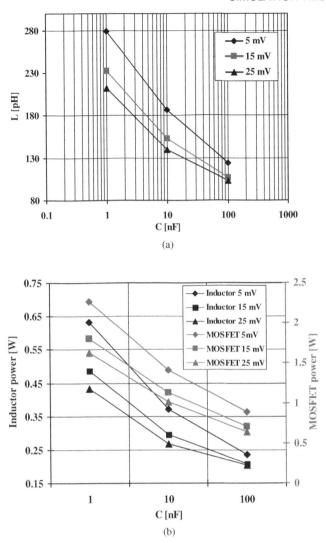

Figure 5.8 Variation of filter inductance and MOSFET- and inductor related-power components of a buck converter with filter capacitance C (1 nF $< C <$ 100 nF) and output voltage ripple ΔV_{DD2} (5 mV $< \Delta V_{DD2} <$ 25 mV). (a) Filter inductance. (b) Total MOSFET- and inductor-related power components

current step from I_{min} to I are shown in Figure 5.9(b) and (c). As shown in Figure 5.9(b), the response time for the buck converter to settle within the allowed 10% voltage window after the microprocessor transitions to the maximum current mode from the idle mode is 87 ns. One solution that provides a stable voltage to the microprocessor until the buck converter output settles within the 10% window is to use several high speed linear regulators distributed around the microprocessor die. These regulators are activated whenever the buck converter output voltage drops below the lower limit of the 10% window. The linear regulator circuits are intrinsically low efficiency voltage converters (for a detailed discussion

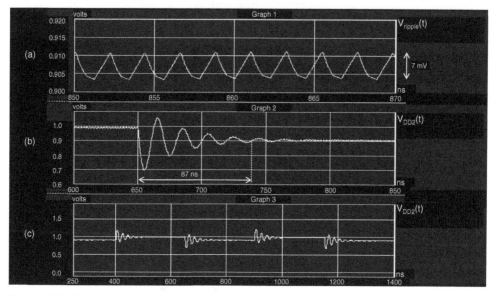

Figure 5.9 Simulation waveforms of a buck converter for $C = 100$ nF. (a) Output voltage ripple $V_{ripple}(t)$. (b) Output response of a buck converter to a change in load current from I_{min} to I. (c) Output response of a buck converter to a step current changing between I_{min} and I

of the linear DC–DC converters see Chapter 4). These large current steps, however, do not occur frequently and the linear regulators are only active for a brief amount of time (a worst case time of 87 ns) until the buck converter output settles within the 10% voltage droop window. The overall impact of these linear regulators on the energy dissipation of the microprocessor is, therefore, small.

5.4 SUMMARY

An analysis of the power characteristics of a standard switching DC–DC converter topology, a buck converter, is provided in this chapter. A parasitic model of a buck converter is presented. With this model, a closed form expression for the total power dissipation of a buck converter is presented. An analysis over a range of design parameters is evaluated, permitting the development of a design space for full integration of active and passive devices on the same die for a target CMOS technology.

Two major challenges for a monolithic switching DC–DC converter are the area occupied by the integrated filter capacitor and the effect of the parasitic impedance characteristics of the integrated inductor on the overall efficiency characteristics of a switching DC–DC converter. A high switching frequency is the key design parameter that enables the integration of a high efficiency buck converter on the same die as a dual-V_{DD} microprocessor.

An optimum switching frequency and inductor current ripple pair that maximizes the efficiency of a buck converter is shown to exist for a target technology. The global maximum efficiency is 92% at a switching frequency of 114 MHz and a current ripple of 9.5 A, assuming an 80 nm CMOS technology. The required filter capacitance and inductance at this operating point are 2083 nF and 104 pH, respectively.

The effects of reducing the filter capacitance due to the tight area constraints on a microprocessor die are examined. An efficiency of 88.4% is described for a circuit operating at a switching frequency of 477 MHz with a filter capacitance of 100 nF. The area occupied by the buck converter is 12.6 mm^2 and is dominated by the area of the integrated filter capacitor. The analytic model for the converter efficiency is within 2.4% of the simulation results at the target design point.

The output voltage ripple can be increased in a fully integrated DC–DC converter, offering the same 10% output voltage droop window as compared to a non-integrated DC–DC converter. The maximum attainable efficiency increases by up to 7.9% as the output voltage ripple is increased from 5 mV to 25 mV. Similarly, the filter inductance and switching frequency required for achieving the maximum efficiency are reduced by 24% and 48.7%, respectively, with the relaxed output voltage ripple constraint.

6 Low-Voltage Swing Monolithic DC–DC Conversion

In a typical non-integrated switching DC–DC converter, significant energy is dissipated in the parasitic impedances of the circuit board interconnect and among the discrete components of the regulator [26], [30], [101]. As the supply current of high performance microprocessors increases with technology scaling, the energy losses of the off-chip power generation and distribution increase, further degrading the efficiency of the DC–DC converters. Integrating both the active and passive devices of a buck converter onto the same die as a dual-V_{DD} microprocessor is presented in Chapter 5 in order to improve efficiency, reduce manufacturing costs, and decrease the number of I/O pads dedicated for power delivery on the microprocessor die. A model is developed and an analysis is presented that describes a design space for full integration of active and passive devices onto the same die as a dual-V_{DD} microprocessor.

As described in Chapter 5, a high switching frequency is the key design parameter that enables the full integration of a high efficiency buck converter. At these high switching frequencies, the energy dissipated in the power MOSFETs and gate drivers dominates the total losses of a DC–DC converter [31]. The efficiency can, therefore, be improved by applying MOSFET power reduction techniques. A low swing MOSFET gate drive technique is described in this chapter that improves the efficiency of a DC–DC converter.

The model described in Chapter 5 provides an accurate representation of the parasitic losses of a full voltage swing buck converter (with an error of less than 2.4% as compared to simulation), but does not provide the flexibility to further optimize the efficiency of the buck converter by varying the driver tapering factors and gate voltages of the power MOSFETs. The independent variables of the buck converter power expressions described in Chapter 5 are the switching frequency f_s and the inductor current ripple Δi. The buck converter model described in Chapter 5 assumes that the PMOS-to-NMOS width ratio within each MOSFET gate driver is two. Similarly, the tapering factor of the MOSFET gate drivers is assumed to be two, assuming a worst case energy efficiency analysis. The signal swing at all of the internal nodes of the buck converter is assumed to be full rail between ground and V_{DD1}. A more comprehensive parasitic model of a buck converter that permits the individual optimization of the gate voltage swings and tapering factors is necessary in order to achieve the objective of this chapter; that is, to design a low voltage swing monolithic DC–DC converter with optimized efficiency for a specific CMOS technology.

A circuit model that permits the optimization of the input and output voltage swing and tapering factor of the power MOSFET gate drivers (in addition to the switching frequency and discrete component values) is described in this chapter. Closed form expressions that characterize the power consumption of a low voltage swing buck converter are presented. The gate voltages and tapering factors of the MOSFETs are included as independent parameters in the model. With the buck converter energy model, a design space is presented which characterizes the integration of both active and passive devices onto the same die. Lowering the voltage swing of the power MOSFET gate drivers is shown to be effective in enhancing the efficiency characteristics of a DC–DC converter.

An efficiency of 84.1% is demonstrated for a voltage conversion from 1.8 V to 0.9 V at the target design point for a full swing DC–DC converter (assuming a 0.18 μm CMOS technology). Expressions for estimating the efficiency of a full swing buck converter are within 0.3% of circuit simulation. The power dissipation of a low swing DC–DC converter is reduced by 27.9% as compared to a full swing DC–DC converter. The maximum efficiency achieved with a low swing DC–DC converter is 88%, 3.9% higher than that achieved with a full swing DC–DC converter.

The chapter is organized as follows. The variable voltage swing and tapering factor DC–DC converter circuit model and closed form expressions characterizing the average power dissipation of a buck converter are presented in Section 6.1. With this model, the efficiency characteristics of a low voltage swing buck converter are analyzed in Section 6.2. A summary of the research results presented in this chapter is provided in Section 6.3.

6.1 CIRCUIT MODEL OF A LOW-VOLTAGE SWING BUCK CONVERTER

A circuit model has been developed to analyze the efficiency characteristics of a low swing buck converter. The circuit model for the parasitic impedances of a buck converter is shown in Figure 6.1.

The power consumed by a buck converter is due to a combination of conduction losses caused by the parasitic resistive impedances and switching losses due to the parasitic capacitive impedances of the circuit components. As discussed in Chapter 5, the power consumed by the pulse width modulation feedback circuit and the integrated filter capacitor is typically small as compared to the power consumed by the power train (the power MOSFETs, MOSFET gate drivers, and the filter inductor). Therefore, only the power dissipation of the power train components is considered in the efficiency analysis.

The MOSFET-related power losses are analyzed in Section 6.1.1. The MOSFET model used during the analysis is discussed in Section 6.1.2. An analysis of the filter inductor-related losses is presented in Section 6.1.3.

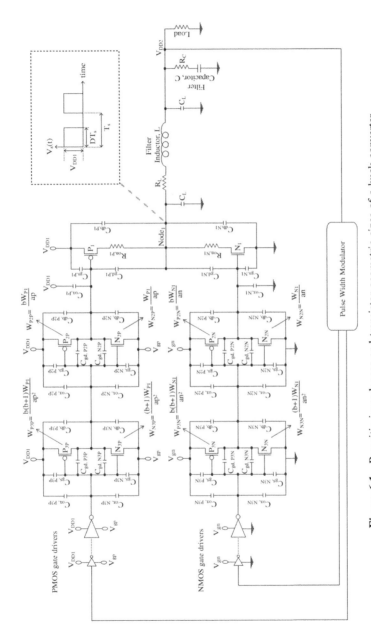

Figure 6.1 Parasitic impedances and transistor geometric sizes of a buck converter

6.1.1 MOSFET Power Dissipation

The total power loss of a MOSFET is a combination of conduction losses and dynamic switching losses. The conduction power is dissipated in the series resistance of the transistors operating in the active region. The dynamic power is dissipated each switching cycle while charging/discharging the gate oxide, gate-to-source/drain overlap, and drain-to-body junction capacitances of the MOSFETs.

As shown in Figure 6.1, the buffers driving P_1 have a ground voltage of V_{gp} where $0 \leq V_{gp} < (V_{DD1} + V_{tp})$. The unit energy (per 1 μm wide power MOSFET) dissipated in the drivers of P_1, assuming $ap > (b + 1)$, is

$$E_{PMOSdrivers} \cong \frac{1}{ap - b - 1}(bC_{OPMOS} + C_{ONMOS})(V_{DD1} - V_{gp})^2, \tag{6.1}$$

$$C_{ONMOS} = C_{ox0NMOS} + 2C_{gd0NMOS} + C_{gs0NMOS} + C_{db0NMOS}, \tag{6.2}$$

$$C_{OPMOS} = C_{ox0PMOS} + 2C_{gd0PMOS} + C_{gs0PMOS} + C_{db0PMOS}, \tag{6.3}$$

where C_{ox0}, C_{gs0}, C_{gd0}, and C_{db0} are the gate oxide, gate-to-source overlap, gate-to-drain overlap, and the drain-to-body junction capacitances, respectively, of a 1 μm wide transistor, ap is the tapering factor of the buffers driving P_1, and b is the ratio of the PMOS to NMOS transistor width within each inverter (see Figure 6.1).

The voltage swing at the gate of P_1 is between V_{gp} and V_{DD1}. The dynamic energy dissipated during a full switching cycle to charge/discharge the parasitic capacitances of a 1 μm wide P-type power transistor is

$$E_{P1} = (C_{ox0PMOS} + C_{gs0PMOS})(V_{DD1} - V_{gp})^2$$
$$+ 2C_{gd0PMOS}\left(-V_{DD1}V_{gp} + V_{DD1}^2 + \frac{V_{gp}^2}{2}\right) + C_{db0PMOS}V_{DD1}^2. \tag{6.4}$$

Combining (6.1), (6.4), and the conduction power dissipated by the effective series resistance of P_1, the total power dissipation related to P_1 is

$$P_{P1TOTAL} = \frac{R_{0PMOS}}{W_{P1}}i_{rmsPMOS}^2 + W_{P1}E_{P1TOTALswitching}f_s, \tag{6.5}$$

$$E_{P1TOTALswitching} = E_{P1} + E_{PMOSdrivers}, \tag{6.6}$$

$$i_{rmsPMOS} = \sqrt{D\left(I^2 + \frac{\Delta i^2}{3}\right)}, \tag{6.7}$$

where R_{0PMOS} is the effective series resistance of a 1 μm wide PMOS transistor, W_{P1} is the width of P_1, f_s is the switching frequency of the buck converter, D is the duty cycle of the signal generated at node$_1$ (see Figure 6.1), I is the DC current supplied to the microprocessor, and Δi is the current ripple of the filter inductor.

As shown in Figure 6.1, the buffers driving N_1 have a supply voltage of V_{gn} ($V_{tn} < V_{gn} \leq V_{DD1}$). The unit energy (per 1 μm wide power MOSFET) dissipated in these buffers, assuming $an > (b + 1)$, is

$$E_{NMOSdrivers} \cong \frac{1}{an - b - 1}(bC_{0PMOS} + C_{0NMOS})V_{gn}^2, \tag{6.8}$$

where an is the tapering factor of the N_1 gate drivers.

The voltage swing at the gate of N_1 is between ground (0 V) and V_{gn}. The dynamic energy dissipated during a full switching cycle to charge/discharge the parasitic capacitances of a 1 μm wide N-type power transistor is

$$E_{N1} = (C_{ox0NMOS} + C_{gs0NMOS} + C_{gd0NMOS})V_{gn}^2 + (C_{gd0NMOS} + C_{db0NMOS})V_{DD1}^2. \tag{6.9}$$

Combining (6.8), (6.9), and the conduction power dissipated in the effective series resistance of N_1, the total power dissipation related to N_1 is

$$P_{N1TOTAL} = \frac{R_{0NMOS}}{W_{N1}}i_{rmsNMOS}^2 + W_{N1}E_{N1TOTALswitching}\,f_s, \tag{6.10}$$

$$E_{N1TOTALswitching} = E_{N1} + E_{NMOSdrivers}, \tag{6.11}$$

$$i_{rmsNMOS} = \sqrt{(1 - D)\left(I^2 + \frac{\Delta i^2}{3}\right)}, \tag{6.12}$$

where R_{0NMOS} is the effective series resistance of a 1 μm wide NMOS transistor and W_{N1} is the width of N_1.

As given by (6.5) and (6.10), increasing the MOSFET transistor width reduces the conduction losses while increasing the switching losses. An optimum MOSFET width, therefore, exists that minimizes the total MOSFET-related power. The optimum transistor width for N_1 and P_1, respectively, is

$$W_{N1opt} = \sqrt{\frac{R_{0NMOS}i_{rmsNMOS}^2}{f_s E_{N1TOTALswitching}}}, \tag{6.13}$$

$$W_{P1opt} = \sqrt{\frac{R_{0PMOS}i_{rmsPMOS}^2}{f_s E_{P1TOTALswitching}}}. \tag{6.14}$$

6.1.2 MOSFET Model

A low swing MOSFET gate drive technique is investigated in this chapter to improve the efficiency of a DC–DC converter. At a reduced gate voltage, the effective series resistance of a MOSFET increases. As discussed in Section 6.1.1, the conduction power dissipated in the series resistance of a power MOSFET constitutes a significant portion of the total MOSFET-related power consumption in a buck converter (half of the total power dissipation of a power

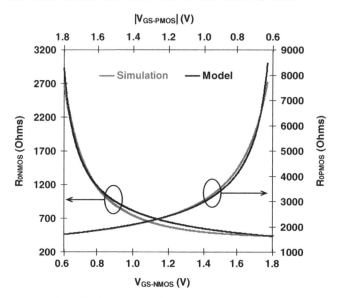

Figure 6.2   Variation of the effective series resistance of 1 µm wide NMOS and PMOS transistors with gate-to-source voltage, V_{GS} ($|V_{DS}| = 0.1$ V)

MOSFET with an optimized transistor width). An accurate MOSFET model is, therefore, required to evaluate the effective series resistance of the MOSFETs at each gate voltage within the range of analysis. The MOSFETs are modeled using the nth power law MOSFET model [91]. As shown in Figure 6.2, the nth power law MOSFET model captures the dependence of the effective series resistance of the MOSFETs on the gate voltages. The worst case error of the model as compared to the simulation data is less than 10%.

6.1.3 Filter Inductor Power Dissipation

Some portion of the total energy consumption of a buck converter occurs due to the series resistance and stray capacitance of the filter inductor. As shown in Chapter 5, the power dissipation in the integrated inductor dominates the total power losses of a buck converter at low switching frequencies.

The integrated filter inductor is a metal slab completely encapsulated by a magnetic material. The magnetic film surrounding the metal is an amorphous cobalt–tantalum–zirconium (CoTaZr) alloy that exhibits a good high frequency response, small hysteresis losses, and can be integrated in a standard high temperature CMOS silicon fabrication process [119], [121].

In the following analysis it is assumed that the parasitic impedances of an integrated inductor scale linearly with the inductance (within the range of analysis) [121]. The total power dissipated in the filter inductor is

$$P_{inductor} = LR_{L0}i_{rms}^2 + \frac{C_{L0}}{L}V_{DD1}^2 f_s, \tag{6.15}$$

$$L = \frac{(V_{DD1} - V_{DD2})D}{2\Delta i \, f_s}, \tag{6.16}$$

where C_{L0} and R_{L0} are, respectively, the parasitic stray capacitance and parasitic series resistance per nH inductance and L is the filter inductance.

6.2 LOW-VOLTAGE SWING BUCK CONVERTER ANALYSIS

The DC–DC converter example described in this chapter provides 1.8 V to 0.9 V conversion while supplying 250 mA per phase DC current to the load in a 0.18 μm CMOS technology. The tapering factor of the P_1 and N_1 drivers is treated as independent variables and ap and an are assumed to be equal ($a = an = ap$). The PMOS-to-NMOS width ratio b within each MOSFET gate driver is assumed to be two. Using the model described in Section 6.2.1, the maximum efficiency attainable for each tapering factor ($8 \leq a \leq 24$) is evaluated.

The switching frequency is the primary design variable used in the analysis. The efficiency of a buck converter is analyzed over the frequency range, 10 MHz $\leq f_s \leq 1$ GHz. At each tapering factor, the maximum attainable efficiency is evaluated over the switching frequency range, varying the circuit configuration. The maximum efficiency circuit configurations as determined by the model are simulated, verifying the circuit operation and performance characteristics.

In the first part of the analysis, the ground voltage of the power PMOS drivers (V_{gp}) and the power supply voltage of the power NMOS drivers (V_{gn}) (see Figure 6.1) are fixed at 0 V and 1.8 V (full swing configuration), respectively. The maximum efficiency attainable with a full swing DC–DC converter is presented in Section 6.2.1. In the second part of the analysis, V_{gp} and V_{gn} are included as independent parameters of the global efficiency optimization process for a low swing configuration. The maximum efficiency attainable with a low swing DC–DC converter is presented in Section 6.2.2.

6.2.1 Full Swing Circuit Analysis for Global Maximum Efficiency

In the first part of the analysis, V_{gp} and V_{gn} (see Figure 6.1) are fixed at 0 V and 1.8 V (full swing configuration), respectively. The maximum efficiency attainable with a full swing buck converter for each tapering factor is shown in Figure 6.3. The global maximum efficiency attainable with a full swing DC–DC converter is 84.1% based on a tapering factor of ten. The switching frequency of the maximum efficiency configuration is 102 MHz. The analytic estimate of the efficiency for the full swing configuration is within 0.3% of the simulations.

The efficiency variation of a buck converter is shown in Figure 6.4 for 10 mA $\leq \Delta i \leq 250$ mA and 10 MHz $\leq f_s \leq 500$ MHz ($a = 10$). The 'z' axis represents the efficiency (%) in Figure 6.4. Similar to the analysis made in Chapter 5 for an 80 nm CMOS technology, the efficiency of a buck converter is characterized by competing inductor and MOSFET losses. At low f_s and Δi, most of the buck converter power is dissipated in the filter inductor. As the switching frequency and current ripple are increased, the inductance is dramatically reduced, lowering the parasitic losses of the inductor. The MOSFET power increases, however, with increasing f_s and Δi. At a certain range of f_s and Δi, the inductor losses dominate the total losses. As shown in Figure 6.4, the efficiency of a buck converter increases with increasing f_s and Δi in the range dominated by the inductor losses. After the peak efficiency is reached, increasing MOSFET losses begin to dominate the total power dissipation of a buck converter. Hence, the efficiency degrades with further increases in f_s and Δi. An optimum switching frequency and inductor current ripple pair exists that maximizes the efficiency of a buck converter. The global

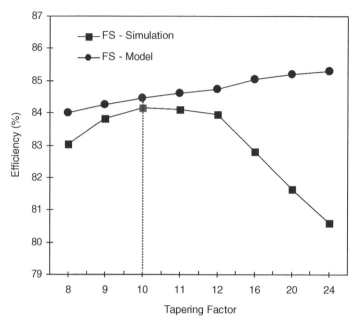

Figure 6.3 The maximum efficiency attainable with a full swing (FS) buck converter circuit for different tapering factors

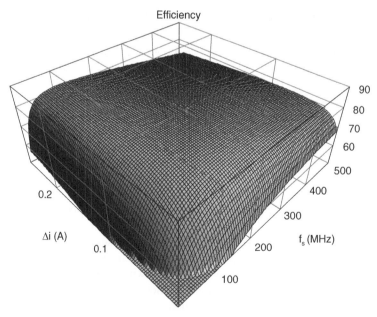

Figure 6.4 Efficiency of a full swing buck converter as a function of the switching frequency (f_s) and inductor current ripple (Δi)

maximum efficiency is 84.1% at a switching frequency of 102 MHz and a current ripple of 250 mA, assuming a 0.18 μm CMOS technology. The required filter inductance at this operating point is 8.8 nH.

In the full swing maximum efficiency circuit configuration, 62% of the total buck converter power is dissipated in the power MOSFETs (P_1 and N_1) and the MOSFET gate driver buffers while 38% of the total power dissipation occurs in the parasitic impedances of the filter inductor. As most of the buck converter energy is dissipated in the MOSFETs, MOSFET-related power reduction techniques can be effective in enhancing the efficiency characteristics of a DC–DC converter.

6.2.2 Low Swing Circuit Analysis for Global Maximum Efficiency

In the second part of the analysis, V_{gp} and V_{gn} are included in the optimization process as independent variables. The effect of reducing the voltage swing of the MOSFET gate driver buffers is explored. For $0 \leq V_{gp} \leq 1.2$ V and 0.5 V $\leq V_{gn} \leq 1.8$ V, an optimal choice of gate voltage is performed at each tapering factor a ($8 \leq a \leq 24$). V_{gp}, V_{gn}, the switching frequency f_s, filter inductance L, and the optimum MOSFET size of the maximum efficiency configurations are determined for each driver tapering factor a. Optimum V_{gp}, V_{gn}, and transistor width (of P_1 and N_1) that maximize efficiency for each a are shown in Figures 6.5 and 6.6, respectively. The optimum circuit configurations obtained from the model have been simulated to verify operation. A comparison of the maximum efficiency attainable by a low swing DC–DC converter and a full swing DC–DC converter for each tapering factor is shown in Figure 6.7.

The total power dissipation of the low swing buck converter is reduced by 27.9% as compared to the full swing maximum efficiency configuration by increasing V_{gp} from 0 to 0.64

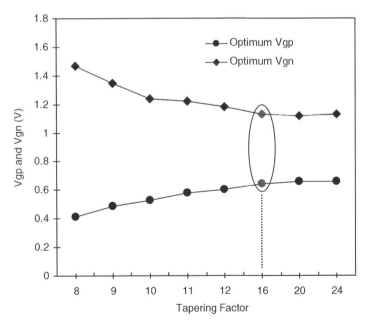

Figure 6.5 Optimum power supply voltage of the power NMOS drivers (V_{gn}) and optimum ground voltage of the power PMOS drivers (V_{gp}) that maximize the efficiency for different tapering factors

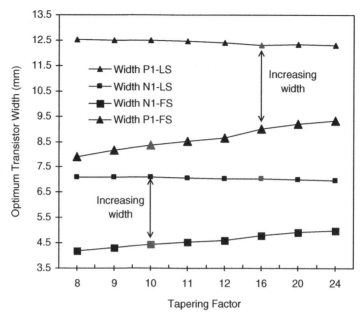

Figure 6.6 A comparison of the optimum width of the power PMOS and NMOS transistors that maximize the efficiency of the full swing (FS) and the low swing (LS) buck converters for different tapering factors

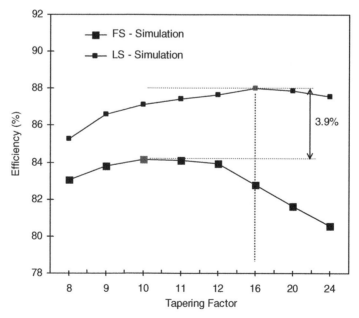

Figure 6.7 A comparison of the maximum efficiency attainable with the low swing (LS) and the full swing (FS) buck converter circuits for different tapering factors

Table 6.1 Efficiency (η) Characteristics of the Full Swing (FS) and Low Swing (LS) DC–DC Converter Circuits Obtained from the Power Model and Simulation ($V_{DD1} = 1.8$ V and $C = 3$ nF)

	FS model	FS simulation	LS simulation
V_{gp} (V)	0	0	0.64
V_{gn} (V)	1.8	1.8	1.13
f_s (MHz)	102	102	102
L (nH)	8.8	8.8	8.8
a	10	10	16
Maximum η (%)	84.4	84.1	88.0
Power reduction	N/A	N/A	27.9%
η difference	+0.3%	0	+3.9%

V and lowering V_{gn} from 1.8 to 1.13 V. As shown in Figure 6.7, the maximum efficiency for a low swing DC–DC converter is 88%, 3.9% higher than achieved with a full swing DC–DC converter. The tapering factor, switching frequency, and filter inductance of the full swing and low swing circuit configurations with the maximum efficiency characteristics are listed in Table 6.1.

The optimal circuit configurations with the highest efficiency characteristics change as the gate voltages are reduced from the full swing voltage. The effective series resistance of a MOSFET is increased while the total dynamic switching energy is decreased with reduced gate voltage. The optimum MOSFET width that minimizes the power dissipation, therefore, increases with a reduced gate voltage swing (as given by (6.13) and (6.14) and as shown in Figure 6.6). As depicted in Figure 6.8, the total transistor width of the power MOSFETs and

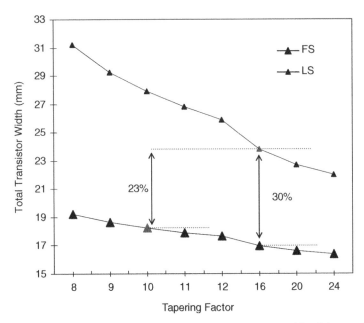

Figure 6.8 A comparison of the total transistor width (including the width of the transistors within the gate drivers) of the low swing (LS) and the full swing (FS) buck converter circuits with the highest efficiency characteristics for different tapering factors

gate drivers for the low swing circuit configuration with the highest efficiency is 23% larger as compared to the full swing circuit with the highest efficiency characteristics.

The described model does not include short-circuit currents in the MOSFET drivers. The model, therefore, produces an efficiency that increases monotonically with increasing a, as shown in Figure 6.3. With increasing tapering factor, the dynamic switching power is reduced while the short-circuit currents increase [31], [120]. At a certain range of a, the dynamic switching energy losses dominate the total losses. As shown in Figure 6.3, the efficiency of a buck converter increases with higher a in the range dominated by switching losses. After the peak efficiency is reached, the increasing short-circuit losses in the power MOSFET gate drivers begin to dominate the total power dissipation of the buck converter. Hence, the efficiency degrades with further increases in a. The optimum tapering factor is 10 and 16 for the full swing and low swing circuits, respectively.

6.3 SUMMARY

A low voltage swing MOSFET gate drive technique is described in this chapter for enhancing the efficiency characteristics of high frequency monolithic DC–DC converters. An analysis of the power characteristics of a standard switching DC–DC converter topology, a buck converter, is provided. Closed form expressions for the total power dissipation of a low swing buck converter are presented. A range of design parameters is evaluated, permitting the development of a design space for full integration of active and passive devices onto the same die for a target CMOS technology.

The effect of reducing the MOSFET gate voltage swings is explored with the circuit model described in this chapter. The optimum gate voltage swing of the power MOSFETs that maximizes efficiency is shown to be lower than the standard full voltage swing. An efficiency of 84.1% is demonstrated for a voltage conversion from 1.8 V to 0.9 V with a full swing monolithic buck converter operating at 102 MHz, assuming a 0.18 μm CMOS technology. It is shown that the power dissipation of a low swing buck converter is reduced by 27.9% as compared to this full swing maximum efficiency configuration by increasing the ground voltage of the power PMOS drivers to 0.64 V and lowering the power supply voltage of the power NMOS drivers to 1.13 V. The maximum efficiency achieved with a low swing DC–DC converter is 88%, 3.9% higher than that achieved with a full swing DC–DC converter.

7 High Input Voltage Step-Down DC–DC Converters

Microprocessors, with higher power consumption and lower supply voltages, demand greater amounts of current from external power supplies with each new technology generation. Voltages significantly higher than current board-level voltages will become necessary to efficiently deliver greater levels of power to future high performance ICs [31]. Distributing power at a higher voltage to the input pads of an IC reduces the supply current. At a reduced supply current, resistive voltage drops and parasitic power dissipation of the off-chip power distribution network is reduced, thereby enhancing the energy efficiency and quality of the distributed voltage [30], [31], [101]. Once the required energy reaches the input pads of a microprocessor, a lower supply voltage for the microprocessor circuitry can be generated by a monolithic DC–DC converter on the same die as the microprocessor, as shown in Figure 7.1.

As discussed in Chapters 5 and 6, monolithic DC–DC conversion on the same die as the load provides several desirable qualities. In a typical non-integrated switching DC–DC converter (as shown in Figure 7.2), significant energy is dissipated in the parasitic impedances of the circuit board interconnect and among the discrete components of the regulator [30], [31]. As microprocessor current demands increase, the energy losses of the off-chip power generation increase, further degrading the efficiency of the DC–DC converters. Integrating both the active and passive devices of a DC–DC converter onto the same die as a microprocessor improves energy efficiency, enhances the quality of the voltage regulation, and decreases the number of I/O pads dedicated for power delivery on the microprocessor die. Furthermore, by employing an IC technology, the reliability of the voltage conversion circuitry can be enhanced, the area can be reduced, and the overall cost of the DC–DC converter can be decreased as compared to a discrete DC–DC converter [30], [31].

Due to the advantages of high voltage power delivery on a circuit board and monolithic DC–DC conversion, next generation low voltage and high power microprocessors are likely to require high input voltage, large step-down DC–DC converters monolithically integrated onto the microprocessor die. The voltage conversion ratios attainable with standard

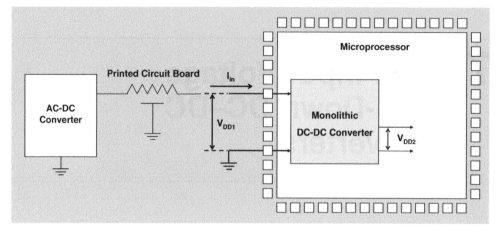

Figure 7.1 High voltage off-chip power delivery and on-chip DC–DC conversion

non-isolated switching DC–DC converter circuits are, however, limited due to MOSFET reliability issues. In a standard buck converter circuit, as shown in Figure 7.2, the input voltage V_{DD1} is limited to the maximum voltage V_{max} that can be directly applied across the terminals of a MOSFET for a specific semiconductor technology. Provided that a DC–DC converter is integrated onto the same die as a microprocessor (fabricated in a low voltage, deep submicrometer CMOS technology), the range of input voltages that can be applied to a standard DC–DC converter circuit is further reduced. A standard non-isolated switching DC–DC converter topology such as the buck converter circuit shown in Figure 7.2 is, therefore, not suitable for future high performance ICs. High efficiency monolithic switching DC–DC converters that can generate very low operating voltages from a significantly higher board-level distribution voltage are highly desirable in scaled nanometer CMOS technologies.

Cascode bridge circuits appropriate for monolithic switching DC–DC converters that provide a high voltage conversion ratio are described in this chapter. The circuits can also be

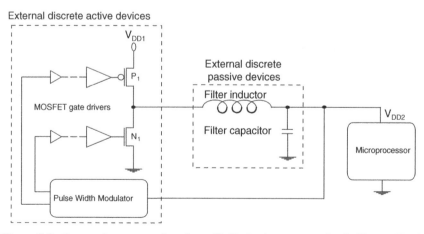

Figure 7.2 Input voltage constraint of an off-chip buck converter circuit ($V_{DD1} \leq V_{max}$)

used as I/O buffers to interface with circuits operating at significantly different voltages. The cascode circuits, when used as part of a voltage regulator, ensure that the voltages across the terminals of all of the MOSFETs in a DC–DC converter are maintained within the limits imposed by the available low voltage CMOS technology. With the cascode circuits described in this chapter, high-to-low non-isolated switching DC–DC converters have been designed based on a 0.18 μm CMOS technology. With the first circuit technique, high-to-low DC–DC converters operating at input voltages up to two times the maximum voltage ($V_{max} = 1.8$ V) are evaluated. An efficiency of 87.8% is demonstrated for a voltage conversion from 3.6 V to 0.9 V while supplying 250 mA of current. The second circuit permits the integration of a DC–DC converter that can operate at input voltages up to three times V_{max}. An efficiency of 79.6% is demonstrated for a voltage conversion from 5.4 V to 0.9 V while supplying 250 mA of DC current.

The chapter is organized as follows. The cascode bridge circuits are described in Section 7.1. The operation and simulation of the voltage converters based on the cascode bridge circuits are described in Section 7.2. A summary of the circuit techniques described in this chapter is provided in Section 7.3.

7.1 CASCODE BRIDGE CIRCUITS

Three cascode bridge circuits are described in this section. The circuits can operate at input voltages higher than the maximum voltage (V_{max}) that can be applied directly across the terminals of a MOSFET in a low voltage nanometer CMOS technology. The cascode circuits can be used for various applications that require high voltage signaling such as I/O drivers and monolithic DC–DC converters. The first cascode bridge circuit allowing input voltages up to twice V_{max} is described in Section 7.1.1. The second cascode bridge circuit can operate at input voltages up to three times V_{max} and is presented in Section 7.1.2. The third circuit further increases the allowable input voltage by up to four times V_{max} and is described in Section 7.1.3.

7.1.1 Cascode Bridge Circuit for Input Voltages up to $2V_{max}$

A cascode buffer operating at an input supply voltage of $V_{DD1} = 2V_{max}$ is shown in Figure 7.3. The cascode buffer circuit generates an output signal swinging between ground and V_{DD1} from an input control signal swinging between ground and V_{max}, while guaranteeing that the voltages applied across the gate-to-source, gate-to-drain, and gate-to-body terminals of the MOSFETs do not exceed the maximum voltage difference, V_{max}, so as to avoid gate oxide breakdown in a CMOS technology. As shown in Figure 7.3, the input supply voltage V_{DD1} can be as high as twice V_{max} while complying with the steady-state voltage constraints across the terminals of all of the MOSFETs.

In Figure 7.3, $V_{DD1} = 2V_{max}$ and $V_{DD3} = V_{max}$. The cascode buffer behaves in the following manner. When the input (node$_7$) transitions low, node$_4$ is pulled up to $2V_{max}$, turning off P$_1$. Node$_6$ is pulled up to V_{max}, turning on N$_1$. The output transitions low through N$_2$ and N$_1$. Node$_1$ is discharged to $V_{max} + |V_{tp}|$. When the input transitions high, node$_6$ is pulled down to ground, cutting off N$_1$. Node$_4$ is pulled down to V_{max}, turning on P$_1$. The output is pulled up to

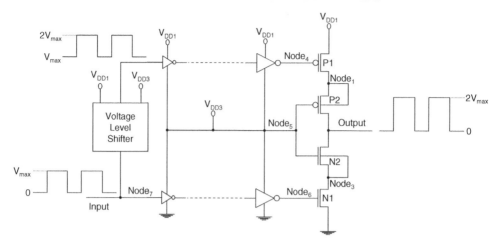

Figure 7.3 Cascode bridge circuit operating at an input supply voltage of $V_{DD1} = 2V_{max}$ ($V_{DD3} = V_{max}$)

$2V_{max}$ through P_1 and P_2. Node$_3$ is charged to $V_{max} - V_{tn}$. Node$_5$ is maintained at V_{max} via V_{DD3}. The source and body terminals of P_2 and N_2 are shorted in order to ensure that the maximum permitted source-to-body and drain-to-body junction reverse bias voltages are not exceeded. With this circuit technique, the voltage differences between the terminals of all of the MOSFETs satisfy the voltage constraints dictated by a low voltage process technology while operating at high input and output voltages up to $2V_{max}$.

7.1.2 Cascode Bridge Circuit for Input Voltages up to $3V_{max}$

The second cascode bridge circuit is shown in Figure 7.4. With this circuit, the input supply voltage V_{DD1} can be as high as three times V_{max} while complying with the steady-state voltage constraints across the terminals of each of the MOSFETs.

In Figure 7.4, $V_{DD1} = 3V_{max}$, $V_{DD3} = 2V_{max}$, and $V_{DD4} = V_{max}$. The number of inverters that drive node$_8$ is even, while the number of inverters that drive node$_6$ and node$_{10}$ is odd. The circuit behaves in the following manner. When the input control signal transitions low, node$_8$ is pulled down to V_{max}. Node$_6$ is pulled up to $3V_{max}$, turning off P_1. Node$_{10}$ is pulled up to V_{max}, turning on N_1. The output transitions low through N_3, N_2, and N_1. Node$_2$ and node$_1$ are discharged to $V_{max} + |V_{tp}|$ and $2V_{max} + |V_{tp}|$, respectively.

When the input control signal transitions high, node$_8$ is pulled up to $2V_{max}$. Node$_{10}$ is pulled down to ground, cutting off N_1. Node$_6$ is pulled down to $2V_{max}$, turning on P_1. The output is pulled up to $3V_{max}$ through P_1, P_2, and P_3. Node$_4$ and node$_5$ are charged to $2V_{max} - V_{tn}$ and $V_{max} - V_{tn}$, respectively. The source and body terminals of P_2, P_3, N_2, and N_3 are shorted to ensure that the maximum permitted source-to-body and drain-to-body junction reverse bias voltages and the maximum permitted gate-to-body voltage are not exceeded. With this circuit technique, the voltage differences between the terminals of all of the MOSFETs satisfy the steady-state voltage constraints dictated by a low voltage process technology while operating at high input and output voltages up to $3V_{max}$.

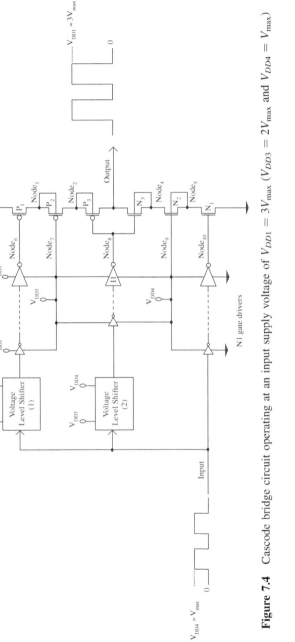

Figure 7.4 Cascode bridge circuit operating at an input supply voltage of $V_{DD1} = 3V_{max}$ ($V_{DD3} = 2V_{max}$ and $V_{DD4} = V_{max}$)

7.1.3 Cascode Bridge Circuit for Input Voltages up to $4V_{max}$

The third driver circuit is shown in Figure 7.5. The circuit operates at input supply voltages ranging up to $4V_{max}$. The driver circuit generates an output signal swinging between ground and $4V_{max}$ from an input signal swinging between ground and V_{max}. The voltages across the nodes of the MOSFETs are adjusted by the driver circuitry ensuring that the voltage-level

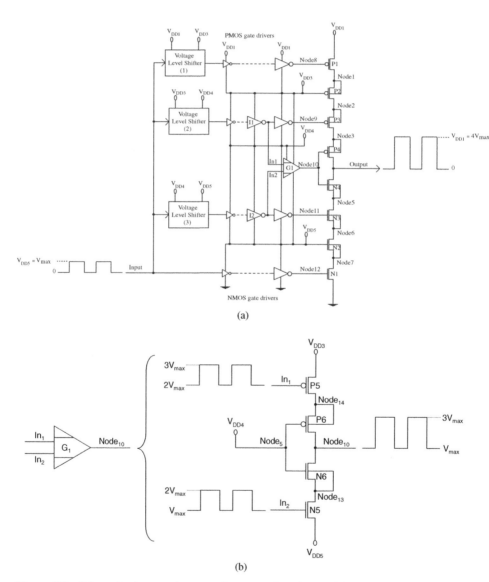

Figure 7.5 Driver circuit operating at an input supply voltage of $4V_{max}$. Output voltage swing is between ground, and $4V_{max}$. $V_{DD1} = 4V_{max}$, $V_{DD3} = 3V_{max}$, $V_{DD4} = 2V_{max}$, and $V_{DD5} = V_{max}$. (a) Cascode bridge circuit. (b) Internal structure of block G_1

constraints are not violated while generating a high voltage swing output signal that is not achievable with standard driver circuits typically available in a low voltage CMOS process.

G_1 generates the necessary control signals for P_4 and N_4 to ensure that the gate-to-source and gate-to-drain voltages of P_4 and N_4 do not exceed the allowable maximum voltage V_{max}. While generating these control signals, the internal circuitry of G_1 ensures that the voltage-level constraints are not violated across the transistors within G_1. When the input transitions high, In_1 and In_2 are pulled down to $2V_{max}$ and V_{max} by the drivers labeled I_1 and I_2, respectively. P_5 is turned on and N_5 is cut off. Node$_{10}$ and node$_9$ are charged to $3V_{max}$. Node$_8$ is discharged to $3V_{max}$, turning on P_1. The output is charged to $4V_{max}$ through P_4, P_3, P_2, and P_1. Node$_5$, node$_6$, and node$_7$ are charged to $3V_{max} - V_{tn}$, $2V_{max} - V_{tn}$, and $V_{max} - V_{tn}$, respectively.

When the input transitions low, In_1 and In_2 are pulled up to $3V_{max}$ and $2V_{max}$ by the drivers labeled I_1 and I_2, respectively. P_5 is cut-off and N_5 is turned on. Node$_{10}$ is discharged to V_{max}. Node$_9$ is discharged to $2V_{max}$. Node$_{11}$ is discharged to V_{max}. Node$_{12}$ is charged to V_{max}, turning on N_1. The output transitions to zero though N_4, N_3, N_2, and N_1. Node$_1$, node$_2$, and node$_3$ are discharged to $3V_{max} + |V_{tp}|$, $2V_{max} + |V_{tp}|$, and $V_{max} + |V_{tp}|$, respectively. The source and body terminals of P_1, P_2, P_3, P_4, N_1, N_2, N_3, and N_4 are shorted to ensure that the maximum permitted source-to-body, and drain-to-body junction reverse bias voltages, and the maximum permitted gate-to-body voltage are not exceeded.

7.2 HIGH INPUT VOLTAGE MONOLITHIC SWITCHING DC–DC CONVERTERS

Step-down DC–DC converters based on the cascode bridge circuits described in Sections 7.1.1 and 7.1.2 are presented in this section. The DC–DC converters have been simulated assuming a $0.18\,\mu m$ CMOS technology. The circuit parameters are optimized to maximize efficiency while satisfying output voltage and current requirements. The efficiency of a DC–DC converter is

$$\eta = 100 \times \frac{P_{load}}{P_{load} + P_{internal}}, \tag{7.1}$$

where P_{load} is the average power delivered to the load and $P_{internal}$ is the average power dissipated in the internal parasitic impedances of a DC–DC converter.

The operation of DC–DC converters operating at input voltages up to $2V_{max}$ and $3V_{max}$ are described in Section 7.2.1. Simulation results characterizing the maximum efficiency circuit configurations for DC–DC converters operating at input supply voltages up to $2V_{max}$ are presented in Section 7.2.2. Simulation results of two voltage converters operating at $V_{DD1} \leq 3V_{max}$ are presented in Section 7.2.3.

7.2.1 Operation of Cascode DC–DC Converters

High-to-low DC–DC converters operating at an input supply voltage of $2V_{max}$ and $3V_{max}$ are shown in Figures 7.6 and 7.7, respectively. The operation of the DC–DC converter circuits behaves in the following manner. The cascode buffers produce an AC signal at node$_2$ (see

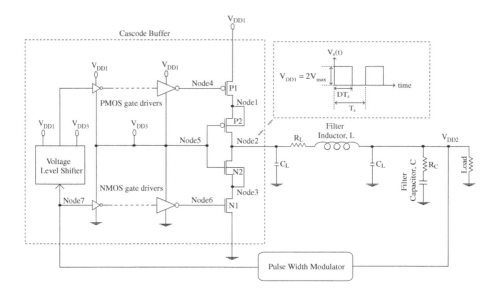

Figure 7.6   DC–DC converter operating at an input supply voltage of $V_{DD1} = 2V_{max}$ ($V_{DD3} = V_{max}$ and $V_{DD2} < V_{DD1}$)

Figure 7.6) and node$_3$ (see Figure 7.7) by a switching action controlled by a pulse width modulator. The AC signal at node$_2$ and node$_3$ is applied to a second order low pass filter composed of an inductor and capacitors. The low pass filter passes the DC component of the AC signal to the output. A small amount of high frequency harmonics (assuming the filter corner frequency is much smaller than the switching frequency f_s of the DC–DC converter) generated by the switching action of the MOSFETs also reaches the output due to the non-ideal characteristics of the output filter.

The output voltage $V_{DD2}(t)$ is

$$V_{DD2}(t) = V_{DD2} + V_{ripple}(t), \tag{7.2}$$

where V_{DD2} is the DC component of the output voltage and $V_{ripple}(t)$ is the voltage ripple waveform observed at the output due to the non-ideal characteristics of the output filter.

The DC component of the output voltage is

$$V_{DD2} = \frac{1}{T_s} \int_0^{T_s} V_s(t)dt = DV_{DD1}, \tag{7.3}$$

where $V_s(t)$ is the AC signal generated at node$_3$ and T_s, D, and V_{DD1} are the period, duty cycle, and amplitude, respectively, of $V_s(t)$. As described by (7.3), any positive DC output voltage less than V_{DD1} can be generated by the DC–DC converters by varying the switching duty cycle D of the pull-up and pull-down network transistors.

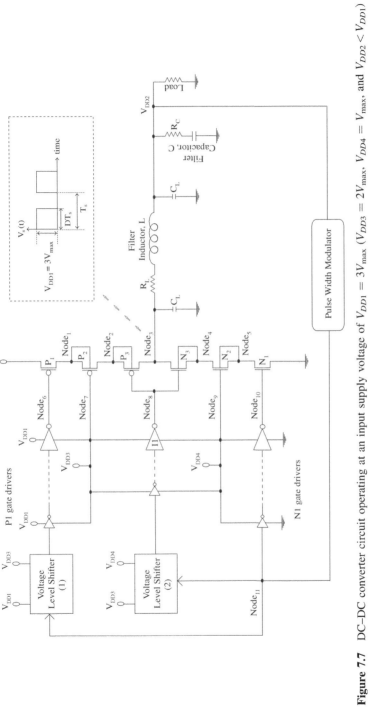

Figure 7.7 DC–DC converter circuit operating at an input supply voltage of $V_{DD1} = 3V_{max}$ ($V_{DD3} = 2V_{max}$, $V_{DD4} = V_{max}$, and $V_{DD2} < V_{DD1}$)

7.2.2 Efficiency Characteristics of DC–DC Converters Operating at Input Voltages up to $2V_{max}$

Three DC–DC converters have been designed based on the first cascode bridge circuit (see Section 7.1.1). The maximum voltage that can be applied across the terminals of a MOSFET (V_{max}) for the targeted 0.18 μm CMOS technology is 1.8 V. The DC–DC converter shown in Figure 7.6 provides 3.6 V ($2V_{max}$) to 0.9 V ($V_{max}/2$) conversion while supplying 250 mA per phase DC current to the load. Two other DC–DC converters have been designed for 2.7 V ($1.5V_{max}$) to 0.9 V ($V_{max}/2$) and 1.8 V (V_{max}) to 0.9 V ($V_{max}/2$) conversion using a similar circuit topology as shown in Figure 7.6.

As listed in Table 7.1, an efficiency of 87.8% is achieved with the DC–DC converter circuit for 3.6 V to 0.9 V conversion. The circuit operates at a switching frequency of 97 MHz. The filter capacitor and inductor of this maximum efficiency circuit configuration are 3 nF and 13.92 nH, respectively.

Efficiencies of 84.8% and 83.5% are observed for the 2.7 V to 0.9 V and 1.8 V to 0.9 V DC–DC converters, respectively. The parasitic energy dissipation within the DC–DC converter increases since the parasitic series resistances of the MOSFETs increase when the input supply voltage V_{DD1} is decreased. The achievable efficiency with the DC–DC converter circuit is, therefore, reduced when the conversion ratio is decreased. Similarly, as listed in Table 7.1, the optimum transistor width that maximizes efficiency increases as the voltage conversion ratio is reduced.

A buck converter circuit has also been designed with the maximum input supply voltage ($V_{DD1} = V_{max} = 1.8$ V) that can be applied to a standard buck converter. For 1.8 V to 0.9 V conversion while supplying 250 mA current, the efficiency attained with a standard buck converter is 4.3% higher than the efficiency achieved with the DC–DC converter described here. The width of the power MOSFETs in a buck converter is also significantly smaller as compared to this DC–DC converter (for a 2 : 1 conversion ratio).

As listed in Table 7.1, the circuit technique offers a similar efficiency while doubling the voltage conversion ratio (as compared to a standard buck converter circuit producing the same output voltage) without creating any MOSFET reliability concerns. The high efficiency achieved for a significantly higher voltage conversion ratio as compared to a standard buck converter circuit can be attributed to a charge recycling mechanism that exists in the cascode buffer circuit. At any time during a state changing transition of the output of the pulse width modulator (independent of the direction of the transition), a portion of the charge required by the inverters to drive $node_6$ originates from the discharging parasitic capacitances of the gate

Table 7.1 Circuit Characteristics of the Optimized Maximum Efficiency Cascode Buffer DC–DC Converters and a Standard Buck Converter For Different Input Supply Voltages

DC–DC converter	Conversion ratio	V_{DD1} (V)	V_{DD2} (V)	V_{DD3} (V)	Max η (%)	f_s (MHz)	C (nF)	L (nH)	W_{NMOS} (mm)	W_{PMOS} (mm)	I_{VDD3} (mA)
Circuit$_1$	4 : 1	3.6	0.9	1.8	87.8	97	3	13.92	6.05	7.25	−2.89
Circuit$_2$	3 : 1	2.7	0.9	1.35	84.8	97	3	12.37	8	11.2	−1.01
Circuit$_3$	2 : 1	1.8	0.9	0.9	83.5	97	3	9.28	11.61	24.39	−0.88
Buck	2 : 1	1.8	0.9	N/A	87.8	97	3	9.28	4.39	9.21	N/A

drivers of P_1 rather than from the power supply V_{DD3}. Most of the charge drawn from V_{DD1} during an output low-to-high transition of a gate driver of P_1 is recycled for use inside the drivers of N_1 before discharged to ground.

As listed in Table 7.1, the average current drawn from V_{DD3} is significantly smaller than the load current. The energy overhead of the additional reference voltage required to properly operate the cascode bridge circuit is therefore small. V_{DD3} can be generated by a simple integrated linear voltage regulator without significantly affecting the overall efficiency of the DC–DC converter. For all three DC–DC converters, the average current supplied by V_{DD3} is negative, meaning that the extra power supply essentially sinks rather than supplies current. The primary purpose of V_{DD3} is therefore to maintain the voltage of node$_5$ at V_{max} rather than supply current to the switching gate driver buffers of N_1.

7.2.3 Efficiency Characteristics of DC–DC Converters Operating at Input Voltages up to $3V_{max}$

Two high-to-low DC–DC converters have been designed based on the second cascode bridge circuit (see Section 7.1.2). The DC–DC converter shown in Figure 7.7 provides 5.4 V ($3V_{max}$) to 0.9 V ($V_{max}/2$) conversion while supplying 250 mA per phase DC current to the load.

Another DC–DC converter circuit has been designed for 4.5 V ($2.5V_{max}$) to 0.9 V ($V_{max}/2$) conversion using a similar circuit topology as shown in Figure 7.7. For the $2.5V_{max}$ to $V_{max}/2$ conversion, V_{DD3} and V_{DD4} are $1.7V_{max}$ and $0.8V_{max}$, respectively, in order to enhance the gate drive of P_1, P_2, and P_3 and to further reduce the voltage stress across the terminals of N_1, N_2, and N_3. The optimized circuit configurations offering the highest efficiency are listed in Table 7.2.

As listed in Table 7.2, an efficiency of 79.6% is achieved with the DC–DC converter circuit for 5.4 V to 0.9 V conversion. The circuit operates at a switching frequency of 97 MHz. The filter capacitor and inductor of this maximum efficiency circuit configuration are 3 nF and 15.5 nH, respectively. Similarly, an efficiency of 79.4% is observed for 4.5 V to 0.9 V conversion. The parasitic energy dissipated within the DC–DC converter is greater as the parasitic series resistance of the MOSFETs increases when the input supply

Table 7.2   Circuit Characteristics of the Maximum Efficiency DC–DC Converters

Conversion ratio	$2.5V_{max} \rightarrow V_{max}/2$ (5:1)	$3V_{max} \rightarrow V_{max}/2$ (6:1)
V_{DD1} (V)	4.5	5.4
V_{DD2} (V)	0.9	0.9
V_{DD3} (V)	3.0	3.6
V_{DD4} (V)	1.5	1.8
I_{out} (mA)	250	250
Max η (%)	79.4	79.6
f_s (MHz)	97	97
C (nF)	3	3
L (nH)	14.8	15.5
W_{N1} (mm)	5.2	5.3
W_{P1} (mm)	7.2	4.8
Total transistor width (mm)	41.3	34.1
I_{VDD3} (µA)	7.6	−178
I_{VDD4} (µA)	205	−186

voltage V_{DD1} is reduced from 5.4 V to 4.5 V. The achievable efficiency with the DC–DC converter circuit is therefore slightly reduced when the conversion ratio is decreased from $6:1$ to $5:1$.

These high efficiencies achieved for such high voltage conversion ratios ($6:1$ and $5:1$) can be attributed to a charge recycling mechanism that exists in the second cascode bridge circuit, similar to the first circuit technique. At any time during a state changing transition of the pulse width modulator output (irrespective of the direction of the transition), a portion of the charge required by the inverters to drive node$_8$ originates from discharging the parasitic capacitances of the gate drivers of P_1 rather than from the power supply V_{DD3}. Similarly, a significant portion of the charge required by the gate drivers of N_1 for a low-to-high output transition originates from discharging the output parasitic capacitances of the gate drivers of P_3 and N_3 rather than from the power supply V_{DD4}. Most of the charge drawn from V_{DD1} during the low-to-high output transition of the buffers driving P_1 is initially recycled for use inside the buffers driving node$_8$. This charge is eventually recycled for use within the buffers driving N_1 before finally being discharged to ground.

As listed in Table 7.2, the average current drawn from V_{DD3} and V_{DD4} is significantly smaller than the load current. The energy overhead of the two extra reference voltages required to properly operate the cascode bridge circuit is, therefore, small. V_{DD3} and V_{DD4} can be generated by simple integrated linear voltage regulators without significantly affecting the overall efficiency of the DC–DC converters. For 5.4 V to 0.9 V conversion, the average current supplied by V_{DD3} and V_{DD4} is negative, meaning that the two additional power supplies essentially sink rather than supply current. For 4.5 V to 0.9 V conversion, the average current supplied by V_{DD3} and V_{DD4} is 7.6 μA and 205 μA, respectively. The primary purpose of V_{DD3} and V_{DD4} is therefore to maintain the voltage at node$_7$ and node$_9$ at $2V_{max}$ and V_{max} ($1.7V_{max}$ and $0.8V_{max}$ for 4.5 V to 0.9 V conversion), respectively, rather than supplying current to the switching gate drivers.

7.3 SUMMARY

Three cascode bridge circuits for use in a monolithic switching DC–DC converter with a high voltage conversion ratio are described in this chapter. The circuits can also be used as I/O buffers to interface circuits operating at significantly different voltages without creating any MOSFET reliability issues due to the high voltage stress. These circuits, when used as part of a voltage regulator, ensure that the voltages across the terminals of all of the MOSFETs in a monolithic DC–DC converter are maintained within the limits imposed by available low voltage CMOS technologies.

High-to-low DC–DC converters have been designed based on the cascode bridge circuits. Reliable operation of the DC–DC converters operating at an input supply voltage up to three times as high as the maximum voltage (V_{max}) that can be directly applied across the terminals of a MOSFET is verified assuming a 0.18 μm CMOS technology. The energy overhead of the DC–DC converter circuit techniques is low due to a charge recycling mechanism operating within the MOSFET gate drivers. An efficiency of 87.8% is demonstrated for a voltage conversion from 3.6 V to 0.9 V while supplying 250 mA of DC current. Similarly, an efficiency of 79.6% is demonstrated for a voltage conversion from 5.4 V to 0.9 V while supplying 250 mA of DC current.

8 Signal Transfer in ICs with Multiple Supply Voltages

As discussed in Chapter 3, in a multiple supply voltage circuit, blocks that are required to operate at high speed utilize a higher supply voltage, while those blocks for which speed is less critical operate at a lower supply voltage. Moreover, due to timing constraints, circuits operating at different supply voltages can exist on the non-critical delay paths in a multiple supply voltage circuit. When a low swing signal drives a CMOS circuit supplied by full rail supply and ground voltages, static DC power is dissipated as the transistors in both the pull-up and pull-down networks are simultaneously turned on. The output voltage swing of a static CMOS gate driven by low swing input signals also degrades. In order to transfer signals among these circuits operating at different voltage levels, specialized voltage interface circuits are required as illustrated in Figure 8.1 [32].

Another application which requires specialized voltage-level converters is the transmission and reception of low swing signals along long interconnects. At each new IC generation, the relative amount of interconnect increases due to the greater number of transistors and the larger die size. In many recent systems, charging and discharging these interconnect lines can require more than 50% of the total power consumed on-chip [125], [126]. In certain programmable logic devices, more than 90% of the total power consumption is due to the interconnect wires [125]. As described in [124]–[126], decreasing the signal voltage swing on the interconnect can significantly lower the power consumed.

A low swing interconnect architecture [126] is shown in Figure 8.2. In this scheme, the circuit blocks operate at a high voltage for high throughput, while a low voltage swing signal is transmitted along the interconnect to decrease the power consumption. Voltage-level converters are placed at the driver and receiver ends of this low swing interconnect architecture to change the voltage swing.

A level converter circuit must consume very low power in order to fully exploit the reduced power achieved by lowering the voltage. In order to not degrade the circuit operating speed, the voltage interface circuit must convert the input signal swing to the desired output signal swing with minimum delay [122], [124]. A simple CMOS interface circuit composed of two cascaded inverters is a standard circuit approach for converting voltage levels [122]–[126].

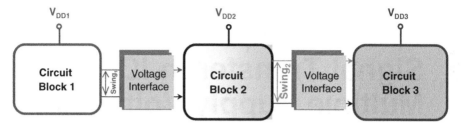

Figure 8.1   Signal transfer between circuit blocks in a multiple supply voltage integrated circuit

This circuit suffers from static power consumption and a non-full rail output voltage swing when converting a low voltage swing input to a high voltage swing output (such as the receiver end shown in Figure 8.2) [32], [122], [123], [126]. Specialized circuits are therefore required to efficiently convert voltage levels.

A bidirectional CMOS voltage interface circuit that drives high capacitive loads to full swing at high speed while consuming no static DC power is presented in this chapter. The propagation delay, power consumption, and power efficiency characteristics of this voltage interface circuit are compared to other interface circuits previously described in the literature [122], [123], [125], [126]. The voltage interface circuit offers significant power savings and lower propagation delay as compared to these circuits.

This chapter is organized as follows. Operation of the interface circuit is described in Section 8.1. Simulation results and a comparison with other converter circuits are presented in Section 8.2. Results from experimental test circuits are presented in Section 8.3. A summary of the key points presented in this chapter is provided in Section 8.4.

8.1   A HIGH-SPEED AND LOW-POWER VOLTAGE INTERFACE CIRCUIT

The voltage interface circuit is shown in Figure 8.3. The circuit provides bidirectional voltage-level conversion. Therefore, without any change in circuit configuration, the interface circuit can be used at both the driver and receiver ends of a low voltage swing circuit architecture (see Figure 8.2) to convert voltage levels from high to low and low to high.

In the interface circuit, P_1 is isolated from the input to minimize both the static power consumption and the propagation delay. As the pull-up and pull-down networks are never simultaneously on, the voltage interface circuit consumes no static DC power while driving high capacitive loads to full swing (V_{DD2}) at high speed.

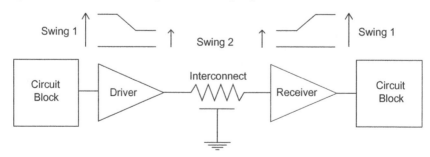

Figure 8.2   Circuit architecture for low swing interconnect

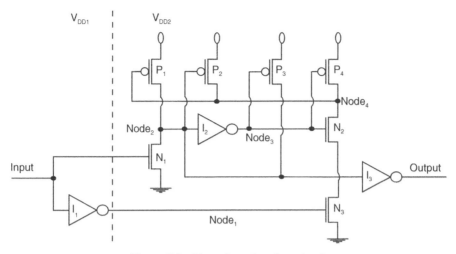

Figure 8.3 The voltage interface circuit

In this circuit, only I_1 is supplied by V_{DD1}. The rest of the circuit (to the right of the demarcation line) is supplied by V_{DD2}. The circuit operates in the following manner. With a $0 \rightarrow 1$ transition at the input, node$_2$ is discharged through N_1. P_2 ensures that P_1 is cut off and I_2 ensures that P_3 is cut off during the output transition, so that the short-circuit power consumption and output transition time are minimized. The output transitions high after node$_2$ becomes sufficiently low. With a $1 \rightarrow 0$ transition at the input, node$_1$ transitions high. Node$_4$ is pulled down to ground through N_2 and N_3 (N_2 is on before the input signal changes). As node$_4$ is discharged to ground, P_1 turns on, charging node$_2$. When node$_2$ is sufficiently high, the output signal transitions low. A negative feedback path exists from node$_3$ to node$_4$ to node$_2$ through I_2, P_4, and P_1. P_3 preserves the output state after P_1 is cut off through the feedback path.

8.2 VOLTAGE INTERFACE CIRCUIT SIMULATION RESULTS

The voltage interface circuit described here is compared to selected voltage interface circuits published in the literature [122], [123], [125], [126]. These circuits are referred to by acronyms derived from the first letters of the last names of the authors who proposed the circuits. The circuit proposed in [122] (SF), the circuit proposed in [123] (CQ), and the circuit described here (KSF) are non-inverting while the asymmetric level converter circuit introduced in [125] (ZGR) and the symmetric level converter circuit introduced in [126] (NIITA) are inverting. To provide a fair comparison, an inverter is added to the output stages of ZGR and NIITA. The output stage inverter of each voltage interface circuit is sized the same.

Simulations are performed for a 0.18 μm CMOS technology. The two voltage levels are 1.8 V and 3.3 V. The simulations have only been carried out for level conversion from low to high since CQ, ZGR, and NIITA have been designed specifically for low-swing-to-high-swing voltage conversion. The input signal applied to each interface circuit is a 1 MHz square wave signal with a 1.8 V swing and a 50% duty cycle. The input to output propagation delay is calculated from 50% of the input swing to 50% of the output swing. The average delay is the arithmetic mean of the high-to-low and low-to-high propagation delays. The average power consumption is calculated for a full cycle of the input waveform.

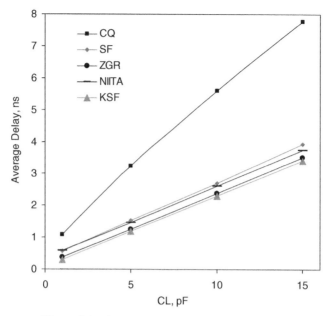

Figure 8.4 Average delay versus load capacitance

Each circuit is optimized to drive a 15 pF load. The load at the output of each interface circuit is swept from 1 pF to 15 pF in order to evaluate the delay and power characteristics. The average propagation delay versus load capacitance characteristics for each of the circuits are shown in Figure 8.4. The average power consumption versus load capacitance is shown in Figure 8.5.

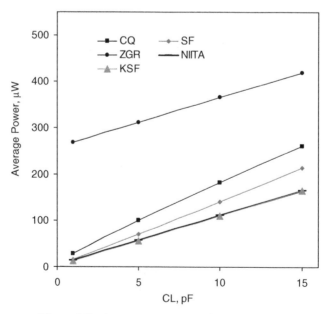

Figure 8.5 Average power versus load capacitance

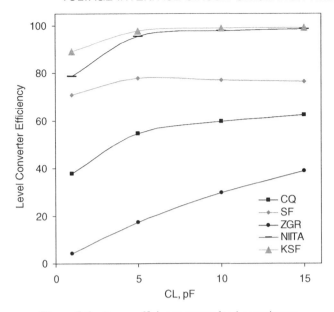

Figure 8.6 Power efficiency versus load capacitance

The voltage interface circuit described here (KSF) exhibits the minimum conversion delay among the target interface circuits. As shown in Figure 8.4, KSF is 3.6 times faster than CQ, 1.9 times faster than SF, 1.2 times faster than ZGR, and 1.9 times faster than NIITA for a 1 pF load capacitance. The propagation delay of ZGR approaches the propagation delay of KSF with increasing load capacitance. However, ZGR displays poor power characteristics as compared to KSF.

The high speed operation of KSF produces no power penalty. Rather, as shown in Figure 8.5, the voltage interface circuit KSF offers significant power savings. KSF reduces the average power consumption by up to 57% as compared to CQ, by up to 24% as compared to SF, by up to 95% as compared to ZGR, and by up to 12% as compared to NIITA.

To better understand the power characteristics of these voltage interface circuits, the power efficiency (defined as the ratio of the power delivered to the load to the total power consumed by the circuit) characteristics are shown in Figure 8.6. The normalized area, maximum frequency of full swing operation (MFSO), and internal power consumption (excluding the power delivered to the load, $C_L = 1$ pF) of each target circuit are listed in Table 8.1. The circuit area is evaluated assuming the area is proportional to the total transistor width. The area of

Table 8.1 Normalized Area, MFSO, and Average Internal Power Consumption of Each Voltage Interface Circuit ($C_L = 1$ pF)

Circuit	Area (normalized)	MFSO (MHz)	Power (µW)
SF [122]	2.8	240	4.5
CQ [123]	2.1	200	17.8
ZGR [125]	1.6	590	257.1
NIITA [126]	1.0	380	2.9
KSF [32]	1.3	610	1.3

each circuit is normalized with respect to the smallest circuit (NIITA). MFSO is defined as the maximum input signal frequency at which a full swing signal is observable at the output for a 1 pF load capacitance.

As shown in Figure 8.6, the internal losses of the KSF circuit are quite small. The power efficiency of KSF ranges from 89.3% to 99.4% as the load is increased from 1 pF to 15 pF. The power efficiency of KSF is 10.3% higher than the power efficiency of NIITA for a 1 pF load. As the load capacitance is increased, the power efficiency of KSF and NIITA both improve and approach each other ($\sim$1% difference) since the internal losses of both circuits become negligible as compared to the power delivered to the load. However, the internal power loss of KSF is significantly lower than the internal power loss of NIITA over the entire range of load capacitances (55% lower for $C_L = 1$ pF and 47% lower for $C_L = 15$ pF).

The power consumed by each circuit increases linearly with the load capacitance (see Figure 8.5). The internal losses of CQ and SF are primarily due to the short-circuit current at the output stage during the output signal transition. As the load capacitance increases, the output transition requires additional time, increasing the short-circuit current. The slopes of the CQ and SF power curves are therefore higher as compared to the other circuits. The worsening short-circuit power loss of SF degrades the efficiency as the load increases above 5 pF (see Figure 8.6). ZGR suffers from significant static DC power loss when the input signal is high, therefore ZGR has the lowest power efficiency (the highest internal power loss).

As listed in Table 8.1, the voltage interface circuit KSF occupies a small amount of area (second smallest) and offers the highest operating frequency range. KSF is operational up to an input frequency of 610 MHz (when driving a 1 pF output load). The MFSO is not directly related to the average delay shown in Figure 8.4 since the MFSO is determined by the longest input-to-output full rail delay (rising or falling) of each circuit.

8.3 EXPERIMENTAL RESULTS

The interface circuit has been fabricated in a 3 μm CMOS technology. A microphotograph of the circuit is shown in Figure 8.7.

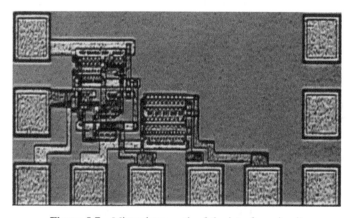

Figure 8.7 Microphotograph of the interface circuit

Table 8.2 Experimentally Measured Test Results

Voltage levels	Output $1 \to 0$ (ns)	Output $0 \to 1$ (ns)
$10\,V \to 5\,V$	190	80
$5\,V \to 10\,V$	120	70

The circuit has been experimentally evaluated with 5 V and 10 V power supplies. To verify the bidirectional operation of the circuit, the circuit has been evaluated for both low-to-high and high-to-low voltage interfaces. The experimental results are listed in Table 8.2. The waveforms obtained from the test circuits are shown in Figure 8.8 (the time axis is 500 ns/ division and the voltage axis is 5 V/division).

The functional operation of the interface circuit has also been experimentally verified. The propagation delays listed in Table 8.2 are higher than the simulation results (see Figure 8.4) due to the voltage level ((1.8 V and 3.3 V) versus (5 V and 10 V)) and feature size (3 μm versus 0.18 μm) differences.

As listed in 8.2, the high-to-low propagation delay is longer than the low-to-high propagation delay for both the $5\,V \to 10\,V$ and $10\,V \to 5\,V$ interfaces. The critical node that determines the output transition time is node$_2$. After a $0 \to 1$ transition at the input, the time to discharge node$_2$ only depends upon the response time of N_1. Alternatively, after a $1 \to 0$ transition at the input, the time to charge node$_2$ depends upon the delay along the path I_1, N_3, N_2, and P_1, thus producing a longer delay.

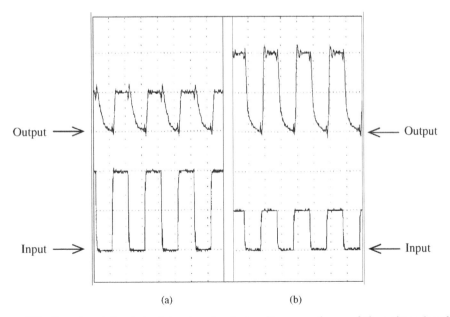

(a)   (b)

Figure 8.8 Experimentally derived input and output voltage waveforms of the voltage interface circuit. (a) $10\,V \to 5\,V$ interface. (b) $5\,V \to 10\,V$ interface

8.4 SUMMARY

A bidirectional CMOS voltage interface circuit for signal transfer between circuits operating at different voltage levels is presented in this chapter. The circuit can also be used at the driving and receiving ends of long interconnect lines so as to lower the power consumption by propagating a smaller voltage swing signal along the line. Up to a 3.6 times delay improvement and up to a 95% power reduction are observed as compared to previously published schemes. The voltage interface circuit operates at high speed while consuming no static DC power.

9 Domino Logic with Variable Threshold Voltage Keeper

Domino logic circuit techniques are extensively applied in high performance micro-processors due to the superior speed and area characteristics of dynamic CMOS circuits as compared to static CMOS circuits [33], [34]. High speed operation of domino logic circuits is primarily due to the lower switching threshold voltage of domino circuits as compared to static gates. This desirable property of a lower switching threshold voltage, however, makes domino logic circuits highly sensitive to noise as compared to static gates. As on-chip noise becomes more severe with technology scaling and increasing operating frequencies, error-free operation of domino logic circuits has become a major challenge [33], [34], [128]–[130].

Threshold voltage reduction accompanies supply voltage scaling, providing enhanced speed while maintaining dynamic power consumption within acceptable levels in each new IC technology generation. Scaling the threshold voltage, however, degrades the noise immunity of domino logic gates [33], [34]. Moreover, exponentially increasing subthreshold leakage currents with reduced threshold voltages have become an important noise source threatening the reliable operation of deep submicrometer (DSM) dynamic circuits [33], [34], [128]–[130].

In a standard domino logic gate, a feedback keeper is employed to maintain the state of the dynamic node against coupling noise, charge sharing, and subthreshold leakage current. The keeper transistor is fully turned on at the beginning of the evaluation phase. Provided that the necessary input combination to discharge the dynamic node is applied, the keeper and pull-down network transistors compete to determine the logical state of the dynamic node. This contention between the keeper and the pull-down network transistors degrades the circuit speed and power characteristics. The keeper transistor is typically sized smaller than the pull-down network transistors in order to minimize the delay and power degradation due to the keeper contention current. A small keeper, however, cannot provide the necessary noise immunity for reliable operation in an increasingly noisy and noise-sensitive on-chip environment [128]–[130]. There is, therefore, a tradeoff between reliability and high speed/energy-efficient operation in domino logic circuits.

A variable threshold voltage keeper circuit technique is described in this chapter for simultaneous power reduction and speed enhancement of domino logic circuits. The current drive of the keeper transistor is dynamically adjusted with the circuit technique. The threshold voltage of the keeper transistor is modified during circuit operation to reduce the contention current without sacrificing noise immunity. The variable threshold voltage keeper circuit technique is shown to enhance circuit evaluation speed by up to 60% while reducing power dissipation by 35% as compared to a standard domino logic circuit. The keeper size can be increased while preserving the same delay or power characteristics as compared to a standard domino circuit since the contention current is reduced with the technique. The domino logic circuit technique offers 14.1%, 8.9%, or 11.9% higher noise immunity under the same delay, power, or power–delay product conditions, respectively, as compared to a standard domino logic circuit technique. Forward body biasing the keeper transistor is also described for improved noise immunity as compared to a standard domino circuit with the same keeper size. It is shown that by applying forward and reverse body bias circuit techniques, the noise immunity and evaluation speed of domino logic circuits are both enhanced.

Challenges in the design of standard domino logic (SD) circuits are reviewed in Section 9.1. The operation of domino logic with a variable threshold voltage keeper (DVTVK) circuit technique is described in Section 9.2. Simulation results characterizing the delay, power, and noise immunity of the DVTVK technique as compared to SD are presented in Section 9.3. The dynamically forward body biased keeper circuit technique for enhanced noise immunity is described in Section 9.4. The research results presented in this chapter are summarized in Section 9.5.

9.1 STANDARD DOMINO (SD) LOGIC CIRCUITS

Performance-critical paths in high performance ICs are often implemented with domino logic circuits. Although domino logic circuit techniques are preferable in high speed circuits, the reliability of domino circuits is seriously degraded in nanometer technologies. The operating principles of domino logic circuits are reviewed in this section. Reliability issues threatening the correct operation of domino logic circuits together with some promising solutions recently proposed in the literature are reviewed. The basic operation of an SD circuit is described in Section 9.1.1. The noise immunity, signal delay, and energy dissipation tradeoffs in domino logic circuits are discussed in Section 9.1.2.

9.1.1 Operation of Standard Domino Logic Circuits

A standard footed domino gate is shown in Figure 9.1(a). Domino circuits behave in the following manner. When the clock signal is low, the domino logic circuit is in the precharge phase. During this phase, the dynamic node is charged to V_{DD1} by the pull-up transistor. The output transitions low, turning on the keeper transistor. When the clock transitions high, the circuit enters the evaluation phase. In this phase, provided that the necessary input combination to discharge the dynamic node is applied, the circuit evaluates and the dynamic node is discharged to ground. If the circuit does not evaluate in the evaluation phase, the high state of the dynamic node is preserved against coupling noise, charge sharing, and subthreshold leakage current by the keeper transistor until the pull-up transistor is turned on at the beginning of the following precharge phase.

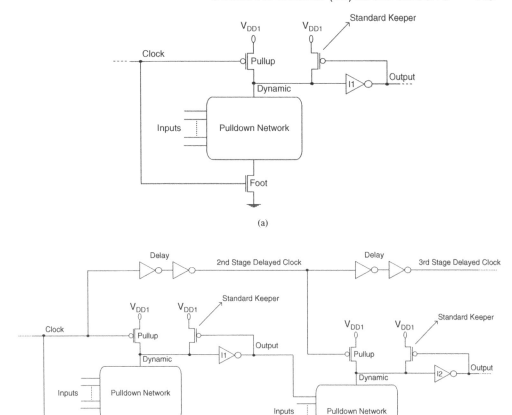

Figure 9.1 Domino gates with standard keeper transistors. (a) Standard footed domino gate. (b) Standard clock-delayed footless domino logic circuit

The foot transistor (see Figure 9.1) controlled by the clock signal divides the operation of a domino logic circuit into two distinct phases independent of the timing of the input signals. The isolation of the pull-down network from ground in the precharge phase eases the relative timing of the input and clock signals in cascaded multi-stage footed domino circuits. If the necessary input combination to discharge the dynamic node is applied during the precharge phase, the pull-down transistors cannot alter the state of the dynamic node as the pull-down path to ground is blocked by the foot transistor.

The foot transistor is a switch in series between the pull-down network and ground. This transistor has a non-zero resistance and parasitic capacitance that degrades the evaluation speed of a domino circuit. The foot transistor is typically sized larger than the pull-down network transistors to minimize this speed degradation. Increasing the size of the foot transistor, however, increases the power consumption since the foot transistor switches every clock cycle. Provided that the clock signal is appropriately delayed, the foot transistors can be omitted in a cascaded multi-stage domino circuit (as shown in Figure 9.1(b)), reducing both the circuit evaluation delay and the power consumption. The clock signal is intentionally delayed from one stage to the next stage in order to ensure that no short-circuit current path from the power supply to ground exists (formed by the pull-up and pull-down network transistors being simultaneously turned on). The clock signal driving a footless domino gate is delayed to transition low only after the previous stage domino gates are all precharged and the inputs to the footless domino gate are all low. Similarly, the inputs to a footless domino gate should transition high only after the clock signal at the gate transitions high and the evaluation phase begins [127]. Although more strict timing of the input and clock signals is required, the overall delay and power characteristics of a footless domino circuit are enhanced as compared to a standard footed domino circuit. Footless domino circuits are, therefore, increasingly popular in high speed ICs [127]. Since the clock signal driving each domino gate is delayed, a multi-stage footless domino circuit is often categorized as a clock-delayed or delayed-precharge domino circuit. As shown in Figure 9.1(b), a first stage domino gate in a multi-stage clock-delayed domino circuit typically has a foot transistor in order to operate with static inputs that may be high during the precharge phase.

9.1.2 Noise Immunity, Delay, and Energy Tradeoffs

As described in Section 9.1.1, the keeper transistor is fully turned on after the output transitions low during the precharge phase. When the clock signal transitions high, the pull-up transistor turns off and the keeper transistor provides the only conductive path between the dynamic node and the power supply, preserving the logical state of the dynamic node in the evaluation phase. Provided that the necessary input combination to discharge the dynamic node is applied during the evaluation phase, the keeper transistor opposes the evaluation of the inputs, degrading the speed and power characteristics of standard domino logic circuits. Current provided by the keeper transistor to charge the dynamic node while the pull-down network transistors are attempting to discharge the dynamic node is called contention current.

The effect of the keeper transistor on the noise immunity, evaluation delay, and power characteristics of domino logic circuits is evaluated assuming a 0.18 μm CMOS technology. The low noise margin (NML) is the noise immunity metric used in this chapter. The NML is defined as

$$NML = V_{IL} - V_{OL}, \tag{9.1}$$

where V_{IL} is the input low voltage defined as the smaller of the DC input voltages on the voltage transfer characteristic (VTC) at which the rate of change of the dynamic node voltage with respect to the input voltage is equal to one (the unity gain point on the VTC). V_{OL} is the output low voltage.

Simulation results for four input standard footless domino AND and OR gates are shown in Figure 9.2. For comparison, simulation results of domino logic circuits without a keeper are also included in Figure 9.2. All of the transistors other than the keeper transistor are sized the same. The effect of the keeper transistor on the circuit delay and noise immunity characteristics varies depending upon the gate input excitation. Simulations of the first group of circuits (NML1, Delay1, and Power1 shown in Figure 9.2) are based on the assumption

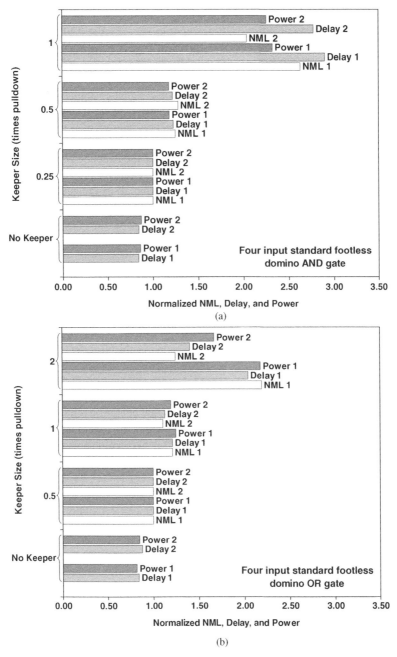

Figure 9.2 Comparison of the normalized noise immunity, evaluation delay, and power characteristics of standard footless domino logic circuits with different keeper sizes. (a) Effect of the increased keeper size on the circuit characteristics of a four input domino AND gate. (b) Effect of the increased keeper size on the circuit characteristics of a four input domino OR gate. NML1, Delay1, and Power1: only one input is excited while the other inputs are either grounded (for the OR gates) or connected to V_{DD} (for the AND gates). NML2, Delay2, and Power2: all four inputs are excited with the same input or noise signal

that the input or noise signals couple only at a single gate input, while the other gate inputs are connected either to ground (for the OR gates) or to V_{DD} (for the AND gates). Additional simulations (NML2, Delay2, and Power2 shown in Figure 9.2) are produced assuming that all of the gate inputs are excited simultaneously by the same input or noise signal.

As shown in Figure 9.2(a), when the input or noise signal is applied to only one input while the other gate inputs are connected to V_{DD} (NML1, Delay1, and Power1 shown in Figure 9.2(a)), the addition of a keeper whose size is a quarter of a pull-down transistor degrades the evaluation speed and power by 16% and 14%, respectively, as compared to a four input domino AND gate without a keeper. By increasing the keeper size from 0.25 to 1, the NML1 is increased by 163%. The increased keeper size, however, also increases the delay and power dissipation by 190% and 132%, respectively. When all of the gate inputs are excited (NML2, Delay2, and Power2 shown in Figure 9.2(a)), the NML2, delay, and power are increased by 104%, 177%, and 125%, respectively, by increasing the keeper size from 0.25 to 1.

When only one input signal is excited while the other three input signals are grounded in a four input domino OR gate, the addition of a keeper half the size of a pull-down network transistor degrades the power and delay by 18% and 16%, respectively, as compared to a standard domino circuit without a keeper (as shown in Figure 9.2(b)). Increasing the keeper size from 0.5 to 2 increases the noise immunity, delay, and power by 119%, 104%, and 118%, respectively. When all of the gate inputs are excited by the same noise or input signal, the effect of the keeper current on both the circuit performance and reliability is reduced. Increasing the keeper size from 0.5 to 2, therefore, improves the NML by only 24%. The delay and power are increased by 40% and 67%, respectively.

As displayed in Figure 9.2, from a circuit performance and energy efficiency point of view, the keeper should be sized as small as possible (or preferably omitted as in earlier domino logic circuits). On the contrary, from a noise immunity and operation reliability point of view, the keeper size should be as large as possible while guaranteeing functionality for a worst case delay input combination. There is, therefore, a tradeoff between high noise immunity and high speed/energy-efficient operation of domino logic gates [128]–[130].

In order to manage these conflicting requirements (a strong keeper for high noise immunity and a weak keeper for high speed), a variable strength keeper scheme (a conditional keeper technique) was first proposed in [128]. Two keeper transistors are employed in the proposed scheme. One of the keeper transistors is sized small in order to reduce the contention current while the other keeper transistor is sized larger for high noise immunity. The larger keeper transistor is conditionally turned on if the dynamic node is not discharged during the evaluation phase. The weak keeper offers limited noise immunity, improving the evaluation speed during the worst case evaluation delay, while the strong keeper offers good robustness to noise and leakage during the rest of the evaluation phase [129]. The primary drawback of this technique is that a delay element and a conditional keeper control circuit are required for each domino gate, increasing the area and energy overhead of the conditional keeper circuits. A similar technique with a single keeper transistor which is cut off at the beginning of the evaluation phase has been proposed in [130]. The dynamic node, without any conductive path to the power supply, floats at the beginning of the evaluation phase. Although the contention current is reduced with the technique proposed in [130], reliable operation cannot be maintained in an increasingly noisy and noise-sensitive on-chip environment. It is assumed with the domino circuit techniques proposed in [128] and [130] that the timing of the clock and input signals driving the domino gates are well known, permitting the worst case evaluation delay to be accurately estimated. The effectiveness of both techniques in reducing the delay and power of domino logic circuits depends upon the accurate estimation of the worst case evaluation delay [129]. Provided that this delay is underestimated, the conditional keeper can be turned on before the evaluation is completed (the dynamic node is fully discharged), causing a contention current on par with the current produced by a standard domino keeper transistor. Alternatively, if the worst case evaluation delay is overestimated, the circuit is exposed to noise with little noise immunity for an extended amount of time, thereby degrading reliability.

A variable threshold voltage keeper circuit technique is described in this chapter for simultaneously reducing power, enhancing speed, and improving noise immunity in domino logic circuits. The current drive of the keeper transistor is adjusted by dynamically body biasing the keeper. The threshold voltage of the keeper transistor is modified during circuit operation to reduce the contention current without sacrificing noise immunity. Similar to the conditional keeper [128] and high speed domino [130] techniques, it is assumed that the worst case evaluation delay of the domino circuits can be accurately predicted. The operation of the DVTVK circuit technique is described in Section 9.2.

9.2 DOMINO LOGIC WITH VARIABLE THRESHOLD VOLTAGE KEEPER (DVTVK)

This circuit technique is introduced in Section 9.2.1. The threshold voltage of the keeper is dynamically modified during circuit operation by changing the body bias voltage of the keeper. Operation of the body bias generator is described in Section 9.2.2.

9.2.1 Variable Threshold Voltage Keeper

A K input domino OR gate based on the variable threshold voltage keeper circuit technique is shown in Figure 9.3. A representative waveform that characterizes the operation of the circuit is shown in Figure 9.4.

The operation of the DVTVK circuit behaves in the following manner. When the clock is low, the pull-up transistor is on and the dynamic node is charged to V_{DD1}. The substrate of the keeper is charged to V_{DD2} ($V_{DD2} > V_{DD1}$) by the body bias generator, increasing the keeper threshold voltage. The value of the high threshold voltage (high-V_t) of the keeper is determined by the reverse body bias voltage ($V_{DD2} - V_{DD1}$) applied to the source-to-substrate p–n junction of the keeper. The current sourced by the high-V_t keeper is reduced, lowering the contention current when the evaluation phase begins. A reduction in the current drive of the keeper does not degrade the noise immunity during precharge as the dynamic node voltage is maintained during this phase by the pull-up transistor rather than by the keeper.

When the clock transitions high (the evaluation phase), the pull-up transistor is cut off and only the high-V_t keeper current contends with the current from the evaluation path transistor(s). Provided that the appropriate input combination that discharges the dynamic node is applied in the evaluation phase, the contention current due to the high-V_t keeper is significantly reduced as compared to SD. After a delay determined by the worst case evaluation delay of the domino gate, the body bias voltage of the keeper is reduced to V_{DD1}, zero biasing the source-to-substrate p–n junction of the keeper. The threshold voltage of

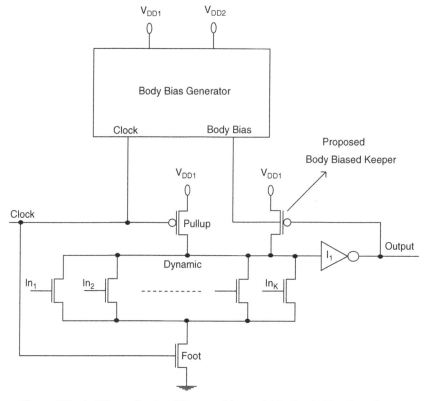

Figure 9.3   A K input domino OR gate with a variable threshold voltage keeper

the keeper is lowered to the zero body bias level, thereby increasing the keeper current. The DVTVK keeper has the same threshold voltage of a standard domino keeper, offering the same noise immunity during the remaining portion of the evaluation phase (assuming the SD and DVTVK keepers are the same size).

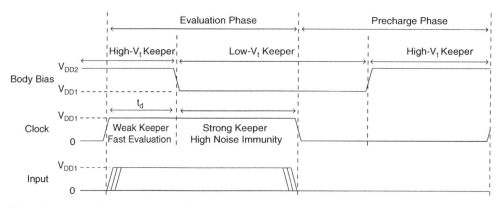

Figure 9.4   Waveforms that characterize the operation of the variable threshold voltage keeper circuit technique

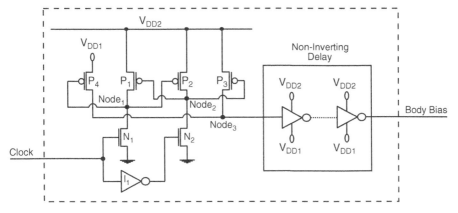

Figure 9.5   Body bias generator circuit

9.2.2 Dynamic Body Bias Generator

The dynamic body bias generator (DBBG) is shown in Figure 9.5. The DBBG produces an output signal swinging between V_{DD1} and V_{DD2} from an input signal swinging between ground and V_{DD1}. The DBBG generates the proper body bias voltages for the keeper with an appropriate delay, ensuring that the contention current is reduced without sacrificing noise immunity.

The operation of the DBBG is controlled by the clock signal that also controls the operational phases of the domino logic circuit. When the clock goes low, node$_2$ is discharged through N_2, turning on P_1 and P_3. P_2 and P_4 are cut off and the body bias voltage is increased to V_{DD2}. When the clock goes high, the domino logic circuit enters the evaluation phase. Node$_1$ is discharged through N_1, turning on P_2 and P_4. P_1 and P_3 are cut off. The voltage at node$_3$ is maintained at V_{DD1} through P_4. During this stage, the DBBG must ensure that the keeper current is increased to the low-V_t level to maintain higher noise immunity if the dynamic node is not discharged by the evaluation path transistors. After a delay determined by the worst case evaluation delay of the domino gate, the body bias voltage is reduced to V_{DD1}. Hence, with a time delay t_d after the clock edge, the threshold voltage of the keeper is reduced to the zero body bias level, increasing the keeper current. During the remaining portion of the evaluation phase, the noise immunity characteristics of the SD and DVTVK circuit techniques are identical.

The DBBG assumes two supply voltages, V_{DD1} and V_{DD2}, where $V_{DD1} < V_{DD2}$. The delay and power savings can be improved by increasing V_{DD2} as compared to V_{DD1}. This change, however, also degrades the noise immunity characteristics of the domino circuit at the beginning of the evaluation phase. The appropriate reverse body bias voltage applied to the keeper is determined by the target delay/power objectives while satisfying the lowest acceptable noise immunity requirements during the worst case evaluation delay of the domino gate. The highest bias voltages that can be applied across the source-to-substrate p–n junction and the gate oxide of a MOSFET for a specific technology are additional factors that determine V_{DD2}.

9.3 SIMULATION RESULTS

As discussed in Section 9.1, the worst case evaluation delay of a wide domino OR gate occurs when only one input is excited while the other inputs are grounded. Similarly, the worst case evaluation delay in a domino gate with stacked pull-down transistors (e.g., an AND–OR or an AND gate) occurs when all of the inputs in the critical pull-down path are excited by the same input signal while all of the other inputs are grounded. The worst case evaluation delay determines the clock speed of a domino circuit while the target clock speed determines the size of the keeper. The speed and power characteristics of the domino logic circuits are evaluated for the set of worst case input vectors. While evaluating the noise immunity, the same noise signal is applied to all of the test circuit inputs as this situation represents the worst case noise condition.

The SD and DVTVK circuit techniques are evaluated for two different test circuits assuming a $0.18\,\mu$m CMOS technology. Simulation results of a multiple-output domino carry generator implemented with the DVTVK circuit technique are presented in Section 9.3.1. The DVTVK circuit technique is also applied to a chain of footless domino OR gates. Simulation results describing clock-delayed domino OR gates (COR) with the DVTVK circuit technique are presented in Section 9.3.2. The effect of gate sizing on the delay and power characteristics of the DVTVK circuit technique is discussed in Section 9.3.3.

9.3.1 Multiple-Output Domino Carry Generator with Variable Threshold Voltage Keeper

A 4-bit multiple-output domino carry generator (CG) implemented with the domino variable threshold voltage keeper (CG-DVTVK) circuit technique is shown in Figure 9.6. A description of the multiple-output domino circuit technique is presented in [133]. The CG circuit has four dynamic nodes. Each node of the CG can be discharged independently by asserting the generate (G) input of the corresponding node. The critical path of the CG circuit is along the N_5–N_9 path. The worst case evaluation delay of the CG occurs while discharging the fourth dynamic node (Dynamic$_4$) through the critical path. During evaluation of the delay and power characteristics, the propagate inputs (P_1–P_4) and C_{in} are asserted while the generate inputs (G_1–G_4) are grounded. While evaluating the noise immunity, all of the inputs are excited by the same noise signal. A 1 GHz clock with a 50% duty cycle is applied to the circuits. All of the common transistors in the SD and DVTVK test circuits are sized the same.

In order to determine an appropriate reverse body bias voltage to be applied to the keeper, the delay, power, power–delay product (PDP), and noise immunity characteristics of CG-DVTVK are evaluated by varying V_{DD2} (for a keeper-to-critical path effective transistor width ratio (KPR) of 2.2). The normalized delay, power, PDP, and NML of CG-DVTVK as compared to the standard domino carry generator (CG-SD) are shown in Figure 9.7. The evaluation delay and power dissipation are reduced by increasing V_{DD2} as compared to V_{DD1}. Increasing V_{DD2}, however, also degrades the noise immunity characteristics of the domino circuit at the beginning of the evaluation phase.

As shown in Figure 9.7, the degradation in noise immunity is 2% for a reverse body bias voltage of 0.3 V while the delay and power savings are 4% and 1%, respectively. By

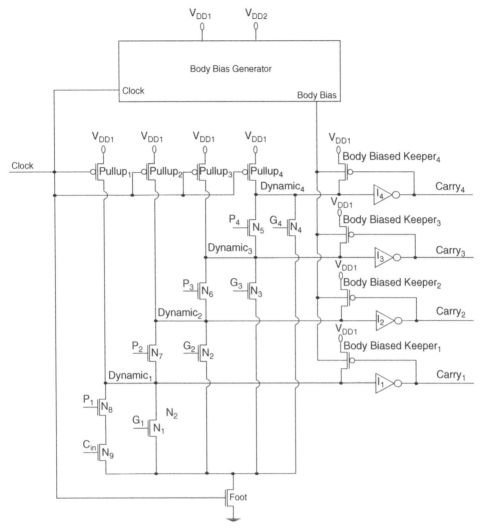

Figure 9.6   A 4-bit multiple-output domino carry generator of a carry lookahead adder implemented with the variable threshold voltage keeper circuit technique. $W_{N2} = 2W_{N1}/3$, $W_{N3} = 2W_{N1}/4$, $W_{N4} = 2W_{N1}/5$, and W_{N5}, W_{N6}, W_{N7}, W_{N8}, $W_{N9} = 2W_{N1}$

increasing the reverse body bias voltage of the keeper transistor to 1.8 V ($V_{DD2} = 3.6$ V), the delay and power savings are increased to 60% and 35%, respectively, while the degradation in noise immunity at the beginning of the evaluation phase increases to 11%. It is assumed that applying a supply voltage of up to 3.6 V to the body bias generator does not create any MOSFET gate oxide-related reliability problems in the target CMOS technology. It is also (arbitrarily) assumed that a degradation of the noise margin by 11% at the beginning of the evaluation phase is acceptable. In the following analysis, V_{DD1} and V_{DD2} are 1.8 V and 3.6 V, respectively.

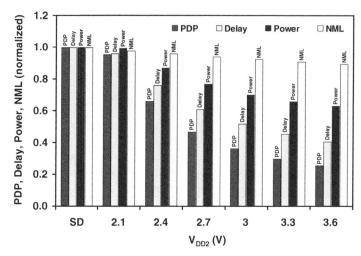

Figure 9.7 Variation of the PDP, delay, power, and NML characteristics of CG-DVTVK with V_{DD2}. Values are normalized to those of an SD carry generator circuit with the same-size transistors (KPR of 2.2)

Simulation results characterizing the delay and power gains achievable with the DVTVK circuit technique for a same-size keeper as compared to SD are analyzed in Section 9.3.1.1. Since the contention current is significantly reduced with the variable threshold voltage keeper circuit technique, the size of the keeper transistor can be increased to improve the noise immunity without degrading the delay and power characteristics as compared to an SD circuit. The improvement in noise immunity offered by the DVTVK technique under the same delay, power, or PDP conditions as compared to SD is presented in Section 9.3.1.2.

9.3.1.1 Improved Delay and Power Characteristics with Comparable Noise Immunity

The keeper width is a multiple of the equivalent width of the pull-down critical path and is varied to evaluate the delay, power, and noise immunity characteristics. The evaluation delay, power, PDP, and NML of the SD and DVTVK circuits as a function of the KPR are shown in Figure 9.8. Provided that the input vector combination that produces the worst case evaluation delay is applied, the fourth dynamic node of the SD circuit cannot be fully discharged during the entire evaluation phase for KPR values above 2.2 due to the high contention current in SD circuits. A KPR of 2.2 is, therefore, the largest value that is considered in this analysis. The gain in delay, power, and PDP achieved by the technique is listed in Table 9.1.

The variable threshold voltage keeper circuit technique is effective for enhancing the evaluation speed of domino logic circuits. As listed in Table 9.1, DVTVK improves the evaluation delay by 60% as compared to SD (for a KPR of 2.2). As shown in Figure 9.8(a), the effectiveness of the technique increases with larger keeper size as the degradation in circuit speed becomes more severe due to increased contention current. The enhancement

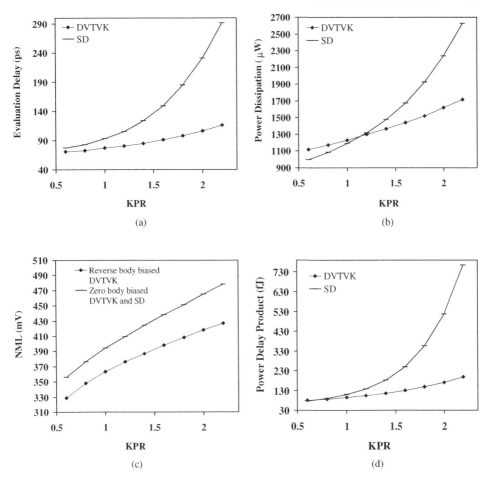

Figure 9.8   SD and DVTVK simulation results for different KPRs. (a) Evaluation delay versus KPR. (b) Power dissipation versus KPR. (c) Noise margin versus KPR. (d) PDP versus KPR

in circuit speed of DVTVK as compared to SD reduces to 8% as the KPR is decreased to 0.6. As shown in Figure 9.8(b), the circuit technique also lowers the power consumption for a wide range of keeper sizes. As listed in Table 9.1, DVTVK reduces the power by 35% as compared to SD (for a KPR of 2.2). As the keeper size is decreased, the effect of the keeper contention current on the evaluation delay and power dissipation becomes smaller.

Table 9.1   A Comparison of the Evaluation Delay, Power Dissipation, PDP, and NML (for Maximum Reverse Body Biased Keeper) of SD and DVTVK Circuit Techniques for KPR of 2.2

	Evaluation delay (ps)	Power (μW)	PDP (fJ)	NML (mV)
SD	291	2625	764	478
DVTVK	116	1717	199	427
Reduction	60%	35%	74%	−11%

The reduction in power, therefore, diminishes with decreasing keeper size. Due to the energy overhead of the dynamic body bias generator circuit, the power consumption of DVTVK is 13% greater than SD when the KPR is reduced to 0.6.

The PDP of the circuits is also illustrated in Figure 9.8 to better compare the effect of the variable threshold voltage keeper circuit technique on circuit performance and energy dissipation. SD has a higher PDP as compared to DVTVK for values of KPR greater than 0.8. As listed in Table 9.1, DVTVK lowers the PDP by 74% as compared to SD for a KPR of 2.2.

Another important metric for domino circuits is the noise immunity. The circuit technique degrades the noise immunity as compared to SD, although only at the beginning of the evaluation phase. This degradation occurs for a brief amount of time until the threshold voltage of the keeper is lowered for increased noise immunity. The time delay (t_d) at the beginning of the evaluation phase, after which the keeper current drive is increased to the low-V_t level, is determined by the worst case evaluation delay of the domino gate. The degradation in noise immunity changes between 8% and 11% under maximum reverse body bias conditions as the KPR is increased from 0.6 to 2.2. As shown in Figure 9.8(c), the noise immunity of DVTVK is identical to the noise immunity of SD whenever a zero body bias is applied to the keeper.

9.3.1.2 Improved Noise Immunity with Comparable Delay or Power Characteristics

The DVTVK circuit technique is shown to offer significant delay and power savings for the same-size keeper as compared to SD. Because of the high contention current in SD circuits, the circuit evaluation delay and power increases significantly with increased keeper size. As explained in Section 9.1, the huge speed and energy penalty incurred to increase the noise immunity in SD circuits is due to the static keeper current drive in the evaluation phase. As shown in Figure 9.8, the NML of SD and zero body biased DVTVK increases by 34% as the KPR is increased from 0.6 to 2.2. The adverse effect of increased keeper size on the delay and power characteristics is significantly lower for DVTVK as compared to SD. As shown in Figure 9.8, the evaluation delay and power dissipation of SD (DVTVK) are increased by 3.8 (1.6) times and 2.6 (1.5) times, respectively, for a 34% noise immunity improvement as the KPR is increased from 0.6 to 2.2. The PDP of SD (DVTVK) increases 10 (2.5) times for a KPR of 2.2 as compared to a KPR of 0.6.

Since the contention current is significantly reduced with the variable threshold voltage keeper technique, the width of the keeper transistor in a DVTVK circuit can be increased without degrading the delay and power characteristics as compared to an SD circuit. DVTVK, therefore, offers higher noise immunity as compared to SD under the same delay, power, or PDP conditions. The KPR of DVTVK is fixed at 2.2 (the highest value considered during the analysis). The SD keeper size is reduced to lower the contention current, offering the same delay, power, or PDP as compared to DVTVK. The improvement in NML of DVTVK as compared to SD (both under the maximum reverse body biased and zero body biased DVTVK keeper conditions) is listed in Table 9.2. The KPR of SD required for the same delay, power dissipation, or PDP characteristics as compared to the DVTVK circuit technique is also listed in the table.

As listed in Table 9.2, the NML of DVTVK is 14.1% higher as compared to SD (zero body biased keeper) when the SD keeper is sized for comparable evaluation speed. Since the keeper

Table 9.2 Achievable Improvement in NML with the DVTVK Circuit Technique as Compared to SD While Maintaining Equal Delay, Power Dissipation, or PDP (KPR of DVTVK is 2.2)

		Noise Margin Improvement as Compared to SD	
	SD-KPR	NML Zero Body Bias	NML Reverse Body Bias
Same delay	1.34	14.1%	1.9%
Same power	1.63	8.9%	−2.7%
Same PDP	1.45	11.9%	0.0%

transistor in the CG-DVTVK circuit is sized 64% larger than the keeper in CG-SD, the noise immunity of CG-DVTVK is higher as compared to CG-SD, even at the beginning of the evaluation phase when the keeper threshold voltage is increased by reverse body biasing the keeper. Under the same power dissipation conditions, the NML of DVTVK with the zero body biased keeper improves by 8.9% as compared to SD. When the PDPs of DVTVK and SD are maintained the same, the DVTVK (with zero body biased keeper) offers an 11.9% higher NML as compared to SD.

9.3.2 Clock-Delayed Domino Logic with Variable Threshold Voltage Keeper

As discussed in Section 9.1, footless domino logic circuits have better speed and power characteristics as compared to footed domino logic circuits. Cascaded footless domino logic circuits, however, require careful timing of the clock and input signals. When the DVTVK circuit technique is applied to a clock-delayed footless domino circuit, the body bias signals should be delayed with respect to the input signals at each footless domino stage. Appropriate timing of the body bias signal is crucial for maximizing the delay and power gains without sacrificing noise immunity with the circuit technique. The DVTVK circuit technique is applied to cascaded footless domino OR gates as shown in Figure 9.9. A three stage chain of eight input domino OR gates with a fan-out of three (COR) is investigated.

A body bias signal that swings between V_{DD1} and V_{DD2} from a clock signal that swings between ground and V_{DD1} is generated in the first stage of a clock-delayed domino circuit. The substrate of the keepers within the domino gates in the following stages are driven by cascaded inverters supplied by V_{DD1} and V_{DD2} (as shown in Figure 9.9). The delay and driving strength of these inverters are adjusted in each domino stage to maintain correct timing of the body bias signals. The clock and body bias signals are delayed at each footless domino stage, maximizing the delay and power gains with the variable threshold voltage keeper circuit technique.

The keeper width is a multiple of the width of a pull-down network transistor (all of the NMOS transistors in a pull-down path are sized the same) and is varied to evaluate the delay, power, and noise immunity characteristics of the chain of domino logic circuits with variable threshold voltage keeper (COR-DVTVK) and the chain of domino logic circuits with standard keeper (COR-SD). A 1 GHz clock with a 50% duty cycle is applied to the circuits. All of the common transistors in the SD and DVTVK test circuits are sized the same. Each domino gate at the third stage drives a 10 fF load. The savings in evaluation

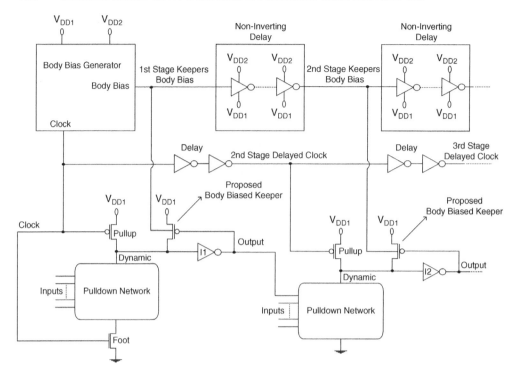

Figure 9.9   Clock-delayed domino logic with the variable threshold voltage keeper circuit technique

delay, power, and PDP of COR-DVTVK as compared to COR-SD for different keeper sizes is listed in Table 9.3.

As listed in Table 9.3, DVTVK improves the evaluation delay, power, and PDP by 6.9%, 0.6%, and 7.5%, respectively, as compared to SD for a KPR of 0.6. The effectiveness of the technique increases with larger keeper size as the degradation in circuit speed and power characteristics becomes more severe due to increased keeper contention current. The

Table 9.3   Delay, Power, and PDP Savings of COR-DVTVK as Compared to COR-SD with Different Keeper Sizes

KPR	Percentage Improvement as Compared to SD			
	Delay	Power	PDP	NML
0.6	6.9	0.6	7.5	−6.1
0.8	9.9	3.2	12.8	−5.9
1.0	12.3	5.7	17.3	−5.9
1.2	15.8	8.8	23.2	−6.0
1.4	19.3	12.7	29.5	−6.0
1.6	23.3	16.8	36.2	−6.1
1.8	28.6	21.9	44.2	−6.2
2.0	35.0	28.5	53.5	−6.5
2.2	43.4	37.2	64.4	−6.4

Table 9.4 Achievable Improvement in NML with the DVTVK Circuit Technique as Compared to SD While Maintaining Equal Delay, Power Dissipation, or PDP (KPR of DVTVK is 2.2)

		Noise Margin Improvement as Compared to SD	
	SD-KPR	NML Zero Body Bias	NML Reverse Body Bias
Same delay	1.45	8.1%	0.0%
Same power	1.61	6.1%	−1.8%
Same PDP	1.52	7.2%	−0.8%

enhancement in circuit speed, power, and PDP of DVTVK as compared to SD is 43.4%, 37.2%, and 64.4%, respectively, for a KPR of 2.2. The degradation in noise immunity (NML) changes between 5.9% and 6.5% as the KPR is varied between 0.6 and 2.2.

Similar to CG-DVTVK, the keeper transistors in a COR-DVTVK circuit can be sized larger, offering higher noise immunity with the same delay and power characteristics as compared to an SD circuit. The keeper transistors of COR-DVTVK and COR-SD are sized for the same delay, power, or PDP characteristics. The improvement in the NML of COR-DVTVK as compared to COR-SD (both under the maximum reverse body biased and zero body biased DVTVK conditions) are listed in Table 9.4. COR-DVTVK offers 8.1% higher noise immunity as compared to SD with the same evaluation speed. The larger size of COR-DVTVK compensates for the reduced gate overdrive ($|V_{gs} - V_{tp}|$) of the keeper transistor at the beginning of the evaluation phase when the keeper is reverse body biased. The noise margins of COR-DVTVK with the reverse body biased keeper and COR-SD for the same evaluation delay are, therefore, equal.

9.3.3 Energy Overhead of the Dynamic Body Bias Generator

It is assumed that each of the CG outputs (in Section 9.3.1) and the third stage footless domino OR gate outputs (in Section 9.3.2) drive a 10 fF load. The transistors in the domino logic circuits have been sized to operate with a 1 GHz clock with a 50% duty cycle. In Figure 9.6, $W_{N1} = 25W_{min}$ and $W_{pull-up} = 8W_{min}$. In Figure 9.9, $W_{pull-down} = 10W_{min}$ and $W_{pull-up} = 9W_{min}$. In the body bias generators, P_1, P_2, P_3, P_4, N_1, N_2, and the transistors within I_1 are minimum sized ($L = L_{min}$ and $W = W_{min}$) while the size and number of inverters have been adjusted to appropriately delay the body bias signals. The DVTVK circuit technique increases the area by 2.3% to 2.8% and 3% to 2.6% as compared to CG-SD and COR-SD, respectively, for a KPR between 0.6 and 2.2. For increasing keeper size, the delay elements (the inverters) are resized to strengthen the body bias signal while most of the transistors forming the DBBG are minimum sized. The energy savings due to the reduced contention current as compared to a standard domino circuit typically exceed the additional energy dissipated by the body bias generator.

The effect of reducing the output load capacitance on the delay and power characteristics of the DVTVK circuit technique is evaluated in this section for a 4-bit multiple-output domino CG and cascaded three stage, eight input COR. The load capacitance is scaled from 10 fF to 2 fF while maintaining a clock frequency of 1 GHz. The savings in the delay, power, and PDP of the CG-DVTVK and COR-DVTVK circuits varies with the load capacitance as shown in Figure 9.10 (KPR of 2.2).

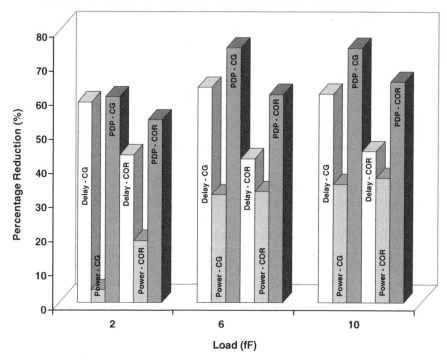

Figure 9.10   Variation of the delay, power, and PDP savings of the CG-DVTVK and COR-DVTVK circuits with the output load capacitance as compared to CG-SD and COR-SD, respectively (KPR of 2.2)

DBBG is used only to drive the substrate of the keeper transistors in the domino logic circuits. Most of the transistors in a DBBG are, therefore, sized minimum even for a high output load capacitance. The energy overhead of DBBG becomes more significant as the pull-up, pull-down, and output inverter transistors of the domino logic circuits are scaled together with the load capacitance. As shown in Figure 9.10, the power savings is, therefore, reduced as the output load capacitance is decreased. The degradation in the power savings of the CG is more significant as compared to COR at small load capacitances. This behavior is explained by the same DBBG being shared by several OR gates in the second and third stages of COR-DVTVK, reducing the overall energy overhead of the DBBG circuits. At high loads, however, the power savings of CG-DVTVK and COR-DVTVK is similar. The enhancement in speed by applying the DVTVK technique is primarily dependent on the relative size of the pull-down network transistors and the keeper. The effectiveness of the DVTVK circuit technique for improving the delay characteristics as compared to SD is, therefore, relatively insensitive to the load capacitance as shown in Figure 9.10 (for the ratio of the same keeper to the pull-down network transistor width).

9.4 DOMINO LOGIC WITH FORWARD AND REVERSE BODY BIASED KEEPER

Reverse body biasing the keeper at the beginning of the evaluation phase is effective for simultaneously improving the speed and power characteristics of domino logic circuits. The keeper transistor is zero body biased after the worst case evaluation delay in order to not sacrifice noise immunity with the variable threshold voltage keeper circuit technique.

Alternatively, forward body biasing the keeper after the worst case evaluation delay is described in this section to improve the noise immunity characteristics as compared to standard domino. The threshold voltage of a forward body biased MOSFET is reduced, increasing the conduction current as compared to a zero body biased transistor with the same physical dimensions. Forward body biasing the keeper, therefore, improves the noise immunity characteristics as compared to an SD circuit with the same keeper size. The DVTVK circuit technique with a forward and reverse body biased keeper is applied to cascaded footless domino OR gates. Simulation results for the COR-DVTVK with a forward body biased keeper are presented in Section 9.4.1. Technology scaling characteristics of the reverse and forward body bias techniques applied to a keeper transistor are discussed in Section 9.4.2.

9.4.1 Clock-Delayed Domino Logic with Forward and Reverse Body Biased Keeper

A three stage chain of eight input domino OR gates with a fan-out of three (COR) is simulated assuming a 0.18 μm CMOS technology. The only difference in the DBBG of the domino circuit with a forward biased keeper is that V_{DD1} (as shown in Figures 9.5 and 9.9) is replaced by a smaller supply voltage V_{DD3} ($V_{DD3} < V_{DD1}$). A body bias signal that swings between V_{DD3} and V_{DD2} from a clock signal that swings between ground and V_{DD1} is generated in the first stage of the clock-delayed domino circuit. The substrate of the keepers within the domino logic gates in the following stages are driven by cascaded inverters supplied by V_{DD3} and V_{DD2}. An eight input footless domino OR gate with a forward body biased keeper is shown in Figure 9.11.

When a keeper transistor is forward body biased, the source-to-body and drain-to-body p–n junctions produce diode currents as illustrated in Figure 9.11. The forward body bias voltage that can be applied to a MOSFET is limited due to these diode currents. The diode current through the drain-to-body p–n junction (I_{diode2}) opposes the drain current (I_{drain}) of the keeper transistor. I_{diode2} attempts to discharge the dynamic node while I_{drain} is charging

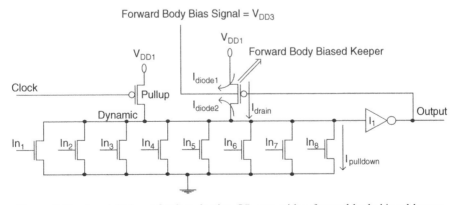

Figure 9.11 An eight input footless domino OR gate with a forward body biased keeper

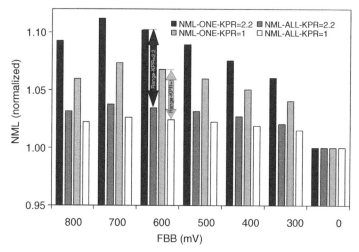

Figure 9.12   Variation of COR-DVTVK noise margins with the FBB for KPRs of 1 and 2.2. The noise margins are normalized to the zero body biased keeper condition. NML-ONE = noise couples to one input while all of the other inputs are grounded. NML-ALL = noise couples to all of the inputs

the node. The drain-to-substrate current, therefore, reduces the net current supplied by the keeper to maintain the state of the dynamic node. The noise margin is greater at forward body bias voltages where the improvement in the keeper drain current due to the reduced threshold voltage dominates the increased drain-to-body junction current. For strongly forward body biased keepers, I_{diode2} lowers (clamps) the voltage of the dynamic node. At room temperature, the DC operating point of the dynamic node when all of the pull-down transistors are cut off (ideal noiseless condition) is reduced by more than 5% for forward body biased (FBB) voltages greater than 700 mV. The noise immunity can, therefore, be reduced, provided that the body diode is strongly turned on at high FBB voltages.

The noise immunity criterion used in this section is similar to the criterion described in [129]. The variation in the noise immunity characteristics of an eight input footless domino OR gate with the body bias voltage applied to the keeper transistor is shown in Figure 9.12, for two different noise coupling scenarios. All of the values are normalized to the standard zero body biased keeper case. As shown in Figure 9.12, increasing the FBB voltage toward 700 mV enhances the noise immunity. For an FBB voltage of 700 mV, the enhancement in noise immunity varies between 3.8% (noise couples to all of the inputs) and 11.2% (noise couples to only one input) as compared to an SD circuit with the same-size transistors (KPR of 2.2). As the FBB voltage is increased beyond 700 mV, the body diodes are strongly turned on, degrading the noise immunity.

An FBB voltage of 700 mV provides the highest enhancement in the noise immunity characteristics at room temperature. For FBB voltages above 600 mV, however, the power overhead of the DVTVK circuit technique significantly increases due to the high diode currents. The variation of the savings in delay, power, and PDP of COR-DVTVK as compared to COR-SD with 500 mV and 600 mV FBB conditions for two different KPR values is illustrated in Figure 9.13. The improvement in delay, power, PDP, and NML of the DVTVK circuit technique as compared to SD for an FBB voltage of 600 mV with two different keeper sizes is listed in Table 9.5.

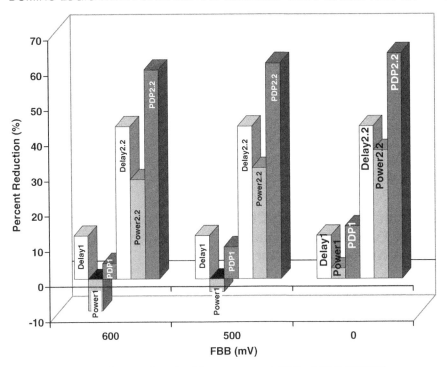

Figure 9.13   Variation of the savings in delay, power, and PDP of COR-DVTVK as compared to COR-SD with an FBB applied to the keeper for two different keeper sizes. {Delay1, Power1, PDP1} → KPR of 1. {Delay2.2, Power2.2, PDP2.2} → KPR of 2.2

The speed enhancement of the DVTVK circuit technique is primarily dependent on the reverse body bias voltage applied to the keeper at the beginning of the evaluation phase. For a V_{DD2} of 3.6 V, therefore, the delay savings of the DVTVK circuit is similar to the delay savings reported in Section 9.3. As shown in Figure 9.13, the improvement in the delay of the DVTVK circuit technique is approximately 43% under the 500 mV and 600 mV FBB conditions.

The power overhead of the DVTVK circuit technique increases when the keeper is forward body biased due to the junction diode currents and the increased voltage swing of the DBBG and keeper substrate (from $V_{DD1} \rightarrow V_{DD2}$ to $V_{DD3} \rightarrow V_{DD2}$). As listed in Table 9.5, the power savings of the DVTVK circuit technique is reduced to 28.3% as the FBB voltage is increased to 600 mV (KPR of 2.2 and load of 10 fF). Similar to the analysis

Table 9.5   Delay, Power, PDP, and NML Savings of COR-DVTVK as Compared to COR-SD (with an FBB Voltage of 0.6 V)

	Improvement (%)				
KPR	Delay	Power	PDP	NML-ALL	NML-ONE
1	12.3	−8.9	4.5	2.4	6.8
2.2	43.4	28.3	59.4	3.5	10.2

described in Section 9.3 for smaller keeper sizes, the effect of the keeper contention current on the evaluation delay and power dissipation is less. The reduction in delay is, therefore, lower and the power savings is smaller with decreased keeper size. As the KPR is reduced to 1, the savings in delay and PDP is reduced to 12.3% and 4.5%, respectively. Since the energy overhead of the DVTVK circuit technique increases when the keeper is forward body biased, the power dissipation of DVTVK is 8.9% higher as compared to SD for a KPR of 1 when the keeper transistor is forward body biased by 600 mV.

For an FBB of 600 mV and KPR of 2.2, the enhancement in noise immunity varies between 3.5% (noise couples to all of the inputs) and 10.2% (noise couples to one input). For a KPR of 1, the range of enhancement in the noise immunity under a 600 mV FBB condition is between 2.4% and 6.8%.

9.4.2 Technology Scaling Characteristics of the Reverse and Forward Body Bias Techniques Applied to a Keeper Transistor

Dynamically adjusting the current drive of the keeper transistors in a domino logic circuit is described in this chapter. The threshold voltage of a keeper transistor is modified during circuit operation by body biasing the keeper transistor. More general schemes have been proposed in the literature for body biasing all of the transistors in order to enhance speed (by lowering the threshold voltage of the transistors), to reduce active power (by lowering both the supply and threshold voltages while maintaining the same speed as compared to a high threshold voltage circuit), to decrease active and standby leakage current (by increasing the threshold voltage of the transistors in the idle portions of a circuit), or to control the within-die and die-to-die threshold voltage variations (by adaptive body biasing) [74], [75], [77], [131], [132]. In a circuit where the body bias voltages of all of the transistors are modified, the power and current demand of the body bias generator can become significant [77]. A DBBG is described in this chapter to drive only the keeper transistors in a domino logic circuit. The power and current demand of the body bias generator for the variable threshold voltage keeper circuit technique is, therefore, small.

Reverse body biasing is typically applied to reduce the subthreshold leakage current (I_{off}) when a circuit is idle [74], [75]. There is an exponential relationship between the subthreshold leakage current and threshold voltage of a MOSFET. Reverse body biasing a transistor increases the threshold voltage, thereby reducing the subthreshold leakage current. Increasing the reverse body bias voltage, however, also increases the band-to-band tunneling current in the source-to-substrate and drain-to-substrate p–n junctions. At high reverse body bias voltages, the increased band-to-band tunneling current becomes comparable to the reduced subthreshold leakage current. There is, therefore, an optimum reverse body bias voltage (limited by the increased band-to-band tunneling currents) that can be applied to a transistor to reduce the total leakage current [74], [75]. Reverse body biasing the keeper transistor is described in order to reduce the active mode conduction current (I_{drain} when the keeper is on) rather than the subthreshold leakage current (I_{off} when the keeper is off). The maximum reverse body bias that can be applied to a keeper transistor is, therefore, typically determined by gate oxide reliability considerations and the domino logic circuit speed and power goals rather than the increased band-to-band tunneling current in the DVTVK circuit technique.

The maximum voltage that can be applied across the gate oxide of a MOSFET is an important factor that limits the reverse body bias voltage. Due to the scaling of the gate oxide thickness, the maximum reverse body bias voltage that can be applied to a keeper can be reduced in future nanometer technology generations. The savings in delay and power of the variable threshold voltage keeper circuit technique as compared to standard domino is reduced at lower keeper reverse body bias voltages as discussed in Section 9.3.1.

The effectiveness of reverse body biasing is reduced with technology scaling due to increasing short-channel and decreasing body effects [74], [75]. Forward body biasing has often been proposed as an alternative to reverse body biasing [74], [131]. Forward body biasing enhances body effect while reducing short-channel effects. FBB is expected to become more effective for controlling the threshold voltage of MOSFETs fabricated in future nanometer process technologies as the supply to threshold voltage ratio decreases with technology scaling [74], [77]. Forward body biasing, however, produces diode currents through the source-to-substrate and drain-to-substrate p–n junctions. These diode currents can become comparable to the drain current of a keeper transistor at low drain-to-source voltages provided the FBB voltage is increased beyond a specific value that is dependent on the junction temperature (700 mV at room temperature). The diode currents degrade the DC operating voltage of the dynamic node even when all of the pull-down transistors are turned off. The diode currents also increase the power overhead of the DVTVK circuit technique. The increased diode currents, therefore, limit the maximum FBB voltage that can be applied to a keeper transistor for enhanced noise immunity.

9.5 SUMMARY

A high speed, low power domino logic circuit technique is described in this chapter. The circuit technique dynamically changes the threshold voltage of the keeper with a specific delay after the beginning of each operational phase (evaluation and precharge) of the domino circuit by varying the body bias voltage of the keeper transistor. The keeper contention current is reduced by increasing the keeper threshold voltage by applying a reverse body bias to the keeper at the beginning of the evaluation phase. Similarly, the degradation in noise immunity of DVTVK as compared to standard domino is avoided by reducing the keeper threshold voltage to the zero body bias level after a delay greater than the worst case evaluation delay of a domino logic circuit. Significant speed enhancements and power reductions are achieved when the keeper is sized for increased noise immunity.

The DVTVK and standard domino circuit techniques are compared in terms of the evaluation delay and power dissipation assuming the DVTVK and standard domino circuits have the same keeper size. The DVTVK technique operates at up to a 60% higher speed while consuming 35% less power as compared to standard domino. DVTVK also reduces the PDP by up to 74% as compared to standard domino. A temporary degradation in the noise immunity of DVTVK as compared to standard domino of less than 11% is observed when the keeper of the DVTVK is reverse body biased.

Since the contention current is significantly reduced with this variable threshold voltage keeper technique, the keeper transistor in a DVTVK circuit can be sized larger, offering higher noise immunity with the same delay and power characteristics as compared to an SD circuit. The DVTVK and SD circuit techniques are compared in terms of the noise immunity that the two circuit techniques offer with the same evaluation delay, power dissipation, or PDP characteristics. For the same evaluation delay characteristics, DVTVK (with a zero biased keeper) offers 14.1% higher noise immunity as compared to standard domino. Under the same power dissipation conditions, DVTVK (with a zero biased keeper) increases the noise immunity by 8.9% as compared to standard domino. Similarly, under the same PDP conditions, DVTVK (with a zero biased keeper) offers 11.9% higher noise immunity as compared to standard domino.

Forward body biasing the keeper transistor is also described to improve the noise immunity as compared to a standard domino circuit with the same keeper size. By applying an FBB of 600 mV to a keeper transistor, the noise immunity is enhanced by up to 10.2%. Dynamically forward and reverse body biasing the keeper transistor simultaneously enhances the noise immunity, evaluation speed, power dissipation, and PDP characteristics of a domino logic circuit.

10 Subthreshold Leakage Current Characteristics of Dynamic Circuits

Subthreshold leakage power is expected to dominate the total power consumption of a CMOS circuit in the near future as depicted in Figure 10.1 [5], [21], [29], [33]–[37]. Energy-efficient circuit techniques aimed at lowering leakage currents are, therefore, highly desirable. The subthreshold leakage current of a domino logic circuit can vary dramatically with the voltage state of the dynamic and output nodes. The dynamic node voltage dependent asymmetry of the subthreshold leakage current characteristics of dual threshold voltage domino gates was first noted in [136]. Based on this asymmetry, several circuit techniques that place dual threshold voltage domino logic circuits into a low leakage state have been proposed in [34], [130], [136], and [142].

A quantitative study of the subthreshold leakage current characteristics of standard low threshold voltage (low-V_t) or dual threshold voltage (dual-V_t) domino logic circuits, however, has to date not been presented in the literature. The node voltage-dependent subthreshold leakage current characteristics of domino logic circuits are examined in this chapter. Different subthreshold leakage current conduction paths which occur during different dynamic and output node voltage states are identified. A discharged dynamic node is preferable for reducing leakage current in a dual-V_t circuit. Alternatively, a charged dynamic node is preferred for lower subthreshold leakage energy in a standard low-V_t domino logic circuit with stacked pull-down devices, such as an AND gate.

Noise immunity issues in dual-V_t domino logic circuits are ignored in [136]. Provided that a dual-V_t CMOS technology is employed, the noise immunity of domino logic circuits can be significantly degraded, affecting the reliability. A brief discussion of noise immunity-related issues in dual-V_t domino circuits is provided in [130]. A dual-V_t domino logic circuit technique based on low-V_t keeper transistors is described in this chapter to maintain a noise immunity similar to standard low-V_t domino logic circuits [130].

A discussion of the effect of dual-V_t CMOS technologies on the noise immunity characteristics of domino logic circuits is provided in this chapter. Two different dual-V_t domino

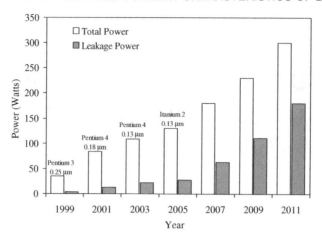

Figure 10.1 Power trends of high performance microprocessors

logic circuit techniques that maintain similar noise immunity as compared to standard low-V_t circuits are evaluated. Both keeper and output inverter sizing is required in a dual-V_t domino logic circuit with a high threshold voltage (high-V_t) keeper transistor in order to provide similar noise immunity as compared to a standard low-V_t domino logic circuit. As an alternative technique, a dual-V_t circuit technique based on low-V_t keeper transistors is also considered in this chapter. Under similar noise immunity conditions as compared to standard low-V_t domino logic circuits, the savings in subthreshold leakage energy achieved by the dual-V_t circuit technique with a high-V_t keeper is 5.7 to 10.9 times higher as compared to the savings offered by the dual-V_t circuit technique with a low-V_t keeper.

Under similar noise immunity conditions, the subthreshold leakage current of dual-V_t domino logic circuits with a high-V_t keeper at a low dynamic node voltage is 224 to 235 times smaller as compared to low-V_t domino logic circuits with a low dynamic node voltage. Alternatively, as compared to low-V_t domino logic circuits with a high dynamic node voltage, the subthreshold leakage current of dual-V_t domino logic circuits with a high-V_t keeper at a low dynamic node voltage is 89 to 3079 times smaller.

The chapter is organized as follows. The node voltage state dependence of the subthreshold leakage current characteristics of different domino logic circuits is described in Section 10.1. The effect of a dual-V_t CMOS technology on the noise immunity characteristics of domino logic circuits is discussed in Section 10.2. The active mode delay and power dissipation of dual-V_t domino logic circuits are presented in Section 10.3. The effect of the difference between the high and low threshold voltages provided in a dual-V_t CMOS technology on the speed and power characteristics of the dual-V_t domino logic circuit technique is evaluated in Section 10.4. The different approaches presented in this chapter are summarized in Section 10.5.

10.1 STATE-DEPENDENT SUBTHRESHOLD LEAKAGE CURRENT CHARACTERISTICS

A dual-V_t domino logic circuit is shown in Figure 10.2. The critical signal transitions that determine the delay of a domino logic circuit occur along the evaluation path. In a dual-V_t

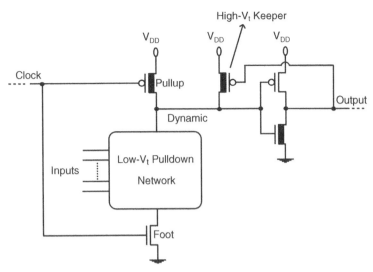

Figure 10.2   A dual-V_t domino logic circuit. The high-V_t transistors are represented by a thick line in the channel region

domino circuit, therefore, all of the transistors that can be activated during the evaluation phase have a low-V_t. The precharge phase transitions are not critical for the speed of a domino logic circuit. In order to exploit the excessive slack of the precharge paths, those transistors that are active during the precharge phase have a high-V_t.

The node voltage dependence of the subthreshold leakage current characteristics of various dual-V_t and low-V_t domino logic circuits is evaluated in this section, assuming a 0.18μm CMOS technology ($V_{tnlow} = |V_{tplow}| = 200\,\text{mV}$, $V_{tnhigh} = |V_{tphigh}| = 500\,\text{mV}$, and $T = 110\,^\circ\text{C}$). The variation of the subthreshold current conduction paths with the node voltages in a low-V_t and dual-V_t domino logic circuit is shown in Figures 10.3 and 10.4, respectively.

Clock gating is an effective method for lowering the dynamic switching power in the unused portions of an integrated circuit. Moreover, when the clock is gated high, the pull-up transistor is turned off, ensuring that no short-circuit current conduction path exists between the power supply and ground (provided that the inputs are high). In this section, therefore, it is assumed that the clock is gated high in an idle domino logic circuit. The dynamic node is cyclically charged every clock period. Therefore, provided that the inputs are low after the clocks are gated, the dynamic node is maintained high during the idle mode, as illustrated in Figures 10.3 and 10.4. Alternatively, provided that the inputs are high after the clocks are gated, the dynamic node is discharged through the pull-down network transistors and the output transitions high, as shown in Figures 10.3 and 10.4. The subthreshold leakage current of a domino logic circuit varies dramatically between these two different states of the dynamic and output nodes, as shown in Figure 10.5.

When the dynamic node voltage is high (the inputs are low), the total subthreshold leakage current of a domino gate is

$$I_{subthreshold\text{-}H} = I_{Leak\text{-}PD} + I_{Leak\text{-}P},\qquad(10.1)$$

where $I_{Leak\text{-}PD}$ and $I_{Leak\text{-}P}$ are the subthreshold leakage currents through the low-V_t pull-down and output inverter pull-up transistors, respectively.

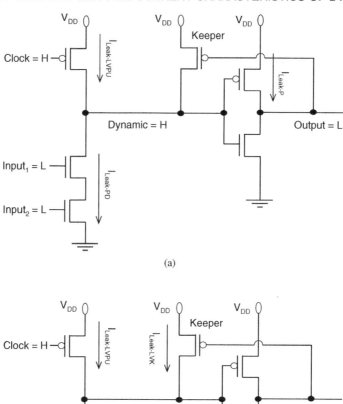

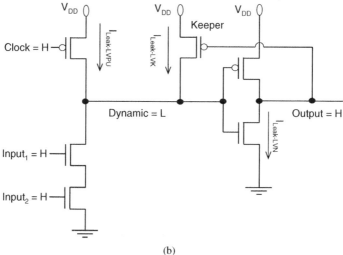

Figure 10.3 Variation of the subthreshold leakage current conduction paths with the state of the dynamic and output nodes in a two input standard low-V_t domino AND gate. (a) High (H) dynamic node voltage state. (b) Low (L) dynamic node voltage state. LVK = low-V_t keeper transistor. LVPU = low-V_t pull-up transistor. LVN = low-V_t NMOS transistor

Alternatively, when the dynamic node voltage is low (the inputs are high), the total subthreshold leakage current of a low-V_t domino gate is

$$I_{subthreshold-L} = I_{Leak-LVPU} + I_{Leak-LVK} + I_{Leak-LVN}, \tag{10.2}$$

where $I_{Leak-LVPU}$, $I_{Leak-LVK}$, and $I_{Leak-LVN}$ are the subthreshold leakage currents through the low-V_t pull-up, keeper, and output inverter pull-down transistors, respectively.

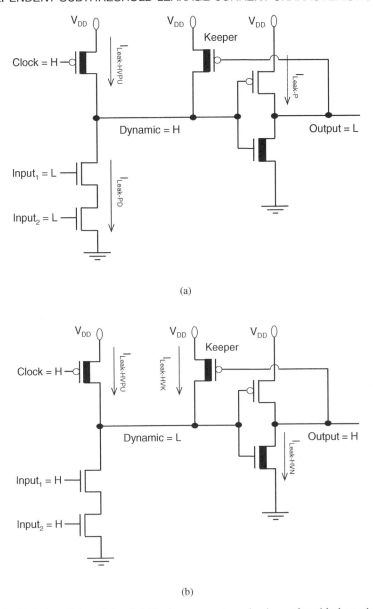

(a)

(b)

Figure 10.4   Variation of the subthreshold leakage current conduction paths with the node voltages in a two input dual-V_t domino AND gate. (a) High (H) dynamic node voltage. (b) Low (L) dynamic node voltage. HVK = high-V_t keeper transistor. HVPU = high-V_t pull-up transistor. HVN = high-V_t NMOS transistor

The subthreshold leakage current through a stack of transistors is orders of magnitude smaller than the subthreshold leakage current through a single transistor [143]. When the inputs are low (the dynamic node is high), $I_{Leak\text{-}PD}$ decreases as more stacked devices are added to the pull-down network. Similarly, as the number of parallel pull-down paths is

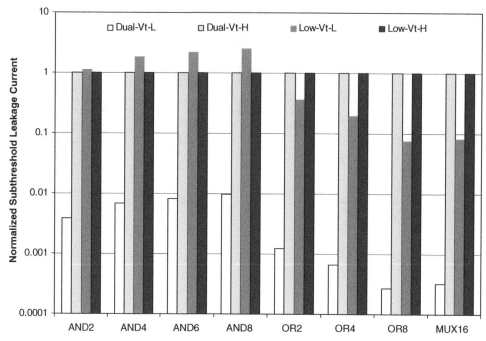

Figure 10.5 Comparison of the subthreshold leakage current of low-V_t and dual-V_t domino logic circuits for the two states of the dynamic node. The leakage current of each gate is normalized to the leakage current of the corresponding low-V_t gate with a high (H) dynamic node voltage. L = low dynamic node voltage. AND2, AND4, AND6, and AND8: two, four, six, and eight input, respectively, domino AND gates. OR2, OR4, and OR8: two, four, and eight input, respectively, domino OR gates. MUX16 = 16-bit domino multiplexer

reduced, $I_{Leak-PD}$ decreases. Alternatively, when the inputs are high (the dynamic node is low), the subthreshold leakage current through the pull-up transistor increases as more stacked devices or parallel discharge paths are added to the pull-down network (due to the increasing width of the pull-up transistor required to drive the increased parasitic capacitance at the dynamic node). $I_{subthreshold-L}$ is higher than $I_{subthreshold-H}$ for a two input low-V_t AND gate. As more stacked devices are added to an AND gate, $I_{subthreshold-H}$ decreases while $I_{subthreshold-L}$ further increases. For a low-V_t domino AND gate, therefore, a high dynamic node voltage is preferred for producing a lower subthreshold leakage current. Alternatively, $I_{subthreshold-H}$ is higher than $I_{subthreshold-L}$ for a two input OR gate. As more parallel discharge paths are added to the pull-down network, both $I_{subthreshold-H}$ and $I_{subthreshold-L}$ increase. Since the increase in $I_{subthreshold-L}$ is smaller than the increase in $I_{subthreshold-H}$, a low dynamic node voltage is preferred for reduced subthreshold leakage current in wide fan-in OR types of gates.

As shown in Figure 10.5, a low dynamic node voltage state produces a 2.8 to 13.2 times smaller subthreshold leakage current as compared to a high dynamic node voltage state in a low-V_t domino circuit with parallel pull-down network paths, such as two, four, and eight input OR gates and a 16-bit multiplexer. Alternatively, in low-V_t domino logic circuits with stacked pull-down network transistors, such as two, four, six, and eight input AND gates, a low dynamic node voltage state produces a 13.3% (AND2) to 153% (AND8) higher subthreshold leakage current as compared to a high dynamic node voltage state.

While the subthreshold leakage current characteristics of low-V_t and dual-V_t circuits are similar for a high dynamic node voltage state, the subthreshold leakage current characteristics of the two circuit techniques are dramatically different for a low dynamic node voltage state. When the inputs are high and the dynamic node voltage is low, the total subthreshold leakage current of a dual-V_t domino gate is

$$I_{subthreshold-L} = I_{Leak-HVPU} + I_{Leak-HVK} + I_{Leak-HVN}, \qquad (10.3)$$

where $I_{Leak-HVPU}$, $I_{Leak-HVK}$, and $I_{Leak-HVN}$ are the subthreshold leakage currents through the high-V_t pull-up, keeper, and output inverter NMOS pull-down transistors, respectively. $I_{Leak-HVPU}$, $I_{Leak-HVK}$, and $I_{Leak-HVN}$ are orders of magnitude smaller than $I_{Leak-LVPU}$, $I_{Leak-LVK}$, and $I_{Leak-LVN}$, respectively. Therefore, provided that the dynamic node is discharged in a domino logic circuit, the subthreshold leakage current can be significantly reduced by employing a dual-V_t CMOS technology, as shown in Figure 10.4. The subthreshold leakage current of dual-V_t domino logic circuits with a low dynamic node voltage is 257 (MUX16) to 293 times (AND2, OR2, and OR4) smaller as compared to low-V_t domino logic circuits with a low dynamic node voltage. Alternatively, as compared to low-V_t domino logic circuits with a high dynamic node voltage, the subthreshold leakage current of dual-V_t domino logic circuits with a low dynamic node voltage is 103 (AND8) to 3719 times (OR8) smaller.

10.2 NOISE IMMUNITY

The noise immunity of low-V_t and dual-V_t domino logic circuits is evaluated in this section. The noise immunity criterion used in this chapter is similar to the criterion described in [129]. The noise margin is the voltage amplitude of the DC noise signal applied to the input that produces a signal with the same amplitude at the output of a domino logic circuit, assuming a 1 GHz clock with a 50% duty cycle.

The degradation in noise immunity due to employing dual-V_t transistors is illustrated in Figure 10.6. The drain current of a high-V_t keeper transistor is reduced as compared to a low-V_t keeper transistor with the same physical size. A dual-V_t domino logic circuit with a high-V_t keeper transistor, therefore, has lower noise immunity as compared to a standard low-V_t domino logic circuit. As illustrated in Figure 10.6, the noise immunity of a dual-V_t domino logic circuit with a high-V_t keeper (HVK) transistor is reduced by 10% (MUX16) to 12.6% (AND2 and OR2) as compared to a low-V_t domino logic circuit with the same size transistors.

The degradation in the noise immunity characteristics in a dual-V_t circuit can be compensated by employing a low-V_t keeper transistor rather than a high-V_t keeper transistor. The noise immunity characteristics of dual-V_t domino logic circuits with a low-V_t keeper (LVK) transistor are illustrated in Figure 10.6. Replacing an HVK transistor with an LVK transistor, as shown in Figure 10.6, is not sufficient to fully compensate for the noise immunity degradation of a dual-V_t domino logic circuit. The noise immunity depends not only on the physical size and threshold voltage of the keeper transistor, but also on the gain of the output inverter. Since the low-V_t NMOS pull-down transistor inside the output inverter of a low-V_t domino logic circuit is replaced by a high-V_t transistor in a dual-V_t domino logic circuit (see Figure 10.4), the high-to-low gain of the output inverter is reduced, further degrading the noise immunity. The noise immunity of a dual-V_t domino logic circuit with an LVK transistor is 3.8% (AND4) to 6.3% (MUX16) lower as compared to a low-V_t domino logic circuit with the same size transistors, as shown in Figure 10.6.

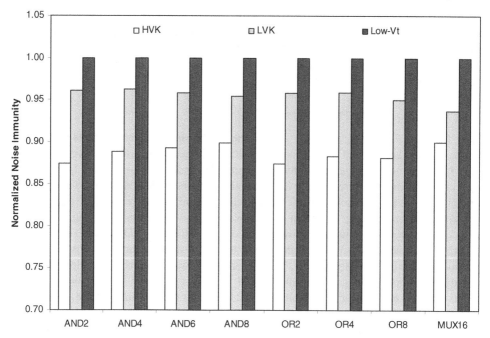

Figure 10.6   Comparison of the noise immunity of low-V_t and dual-V_t domino logic circuits with the same size transistors. The noise margin of each gate is normalized to the noise margin of the corresponding low-V_t gate. HVK = high-V_t keeper. LVK = low-V_t keeper

One circuit technique for maintaining the same noise immunity as compared to a low-V_t circuit is to employ an LVK while increasing the size of the high-V_t pull-down transistor within the output inverter, thereby enhancing the high-to-low output gain and noise immunity. Another circuit technique to compensate the degradation in noise immunity is to increase the width of the HVK and the high-V_t NMOS pull-down transistor within the output inverter. In this section, both techniques are applied to domino logic circuits to enhance noise immunity. A comparison of the subthreshold leakage current characteristics of dual-V_t domino with an HVK (dual-V_t–HVK), dual-V_t domino with an LVK (dual-V_t–LVK), and low-V_t domino logic circuit techniques while providing similar noise immunity characteristics is shown in Figure 10.7.

Increasing the physical size of the keeper and the pull-down transistor within the output inverter increases both $I_{Leak\text{-}HVK}$ and $I_{Leak\text{-}HVN}$, thereby degrading the reduction in subthreshold leakage current achieved at a low dynamic node voltage state. As shown in Figure 10.7, under similar noise immunity conditions, the subthreshold leakage current of dual-V_t domino logic circuits with an HVK and a low dynamic node voltage is 224 (AND8) to 235 times (MUX16) smaller as compared to low-V_t domino logic circuits with a low dynamic node voltage. Alternatively, as compared to low-V_t domino logic circuits with a high dynamic node voltage, the subthreshold leakage current of dual-V_t domino logic circuits with an HVK and a low dynamic node voltage is 89 (AND8) to 3079 times (OR8) smaller.

For the high voltage state of the dynamic node, the subthreshold leakage current characteristics of dual-V_t domino circuits are similar to low-V_t circuits. When the dynamic node voltage is low, the subthreshold leakage current of a dual-V_t domino logic circuit is

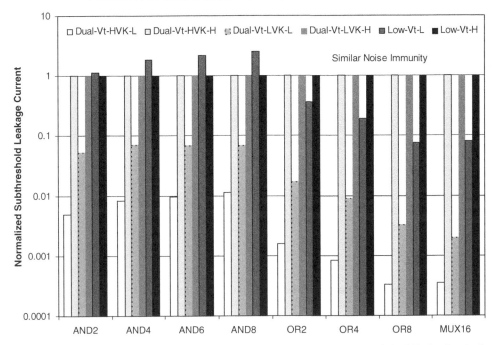

Figure 10.7 Comparison of the subthreshold leakage current of low-V_t and dual-V_t domino logic circuits for the two states of the dynamic node (under similar noise immunity conditions). The leakage current of each gate is normalized to the leakage current of the corresponding low-V_t gate with a high dynamic node voltage (H). L = low dynamic node voltage. Dual-V_t-HVK = dual-V_t domino with high-V_t keeper. Dual-V_t-LVK = dual-V_t domino with low-V_t keeper. Low-V_t = standard low-V_t domino circuit

significantly increased, provided that an LVK rather than an HVK transistor is employed. The subthreshold leakage current conduction paths within a dual-V_t domino logic circuit with a low-V_t keeper at a low dynamic node voltage are shown in Figure 10.8. When the dynamic node voltage is low, the total subthreshold leakage current of a dual-V_t domino gate with an LVK is

$$I_{subthreshold-L} = I_{Leak-HVPU} + I_{Leak-LVK} + I_{Leak-HVN}, \qquad (10.4)$$

where $I_{Leak-LVK}$ is the subthreshold leakage current through an LVK transistor.

Under similar noise immunity conditions, the subthreshold leakage current of dual-V_t domino logic circuits with an LVK at a low dynamic node voltage is 21 (AND2, OR2, and OR4) to 41 times (MUX16) smaller as compared to low-V_t domino logic circuits with a low dynamic node voltage. Alternatively, as compared to low-V_t domino logic circuits with a high dynamic node voltage, the subthreshold leakage current of dual-V_t domino logic circuits with an LVK at a low dynamic node voltage is 14 (AND4, AND6, and AND8) to 503 times (MUX16) smaller. Since $I_{Leak-LVK}$ is higher than $I_{Leak-HVK}$, the subthreshold leakage current of dual-V_t domino circuits with an LVK is 5.7 (MUX16) to 10.9 times (OR4) higher than the subthreshold leakage current of dual-V_t domino logic circuits with an HVK, under similar noise immunity conditions.

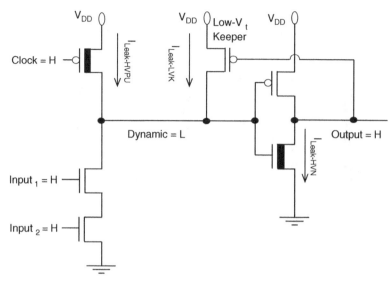

Figure 10.8   Subthreshold leakage current conduction paths for the low (L) voltage state of the dynamic node in a dual-V_t domino AND gate with a low-V_t keeper. LVK = low-V_t keeper transistor. HVPU = high-V_t pull-up transistor. HVN = high-V_t NMOS transistor

10.3 POWER AND DELAY CHARACTERISTICS IN THE ACTIVE MODE

The evaluation delay, precharge delay, and power consumption of dual-V_t and low-V_t domino logic circuits are evaluated in this section. The evaluation and precharge delay of example domino circuits are shown in Figures 10.9 and 10.10, respectively. The power consumption characteristics of dual-V_t and low-V_t domino logic circuits are illustrated in Figure 10.11.

As shown in Figures 10.9 and 10.11, dual-V_t domino logic circuits have reduced evaluation delay and power consumption as compared to low-V_t domino logic circuits with the same size transistors. The enhancement in the delay and power characteristics is primarily due to the reduced contention current [33] of an HVK transistor and the increased low-to-high gain of the output inverter.

In dual-V_t circuits, provided that the HVK and the output inverter are sized to maintain a similar noise immunity (SNI) as compared to low-V_t circuits, except for the eight input AND gate, the evaluation delay is greater as compared to the low-V_t circuits. The degradation in the evaluation speed is less than 11.8% (OR4). The increase in the precharge delay of the dual-V_t domino circuits with HVKs is less than 23.4% (OR8) as compared to the low-V_t circuits. The increase in the active mode power consumption is less than 5.1% (AND2). For the six and eight input AND gates and the 16-bit multiplexer, the dual-V_t domino logic circuit technique reduces the power consumed in both the active and standby modes while providing a similar noise immunity as compared to the low-V_t circuit technique.

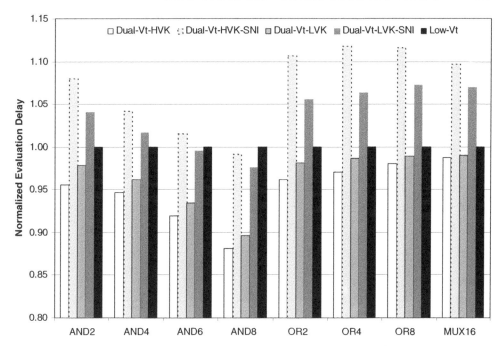

Figure 10.9 Comparison of the evaluation delay of domino logic circuits. The evaluation delay of each gate is normalized to the delay of the corresponding low-V_t gate. SNI = similar noise immunity

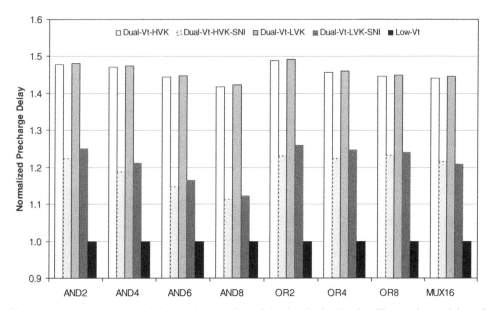

Figure 10.10 Comparison of the precharge delay of domino logic circuits. The precharge delay of each gate is normalized to the precharge delay of the corresponding low-V_t gate. SNI = similar noise immunity

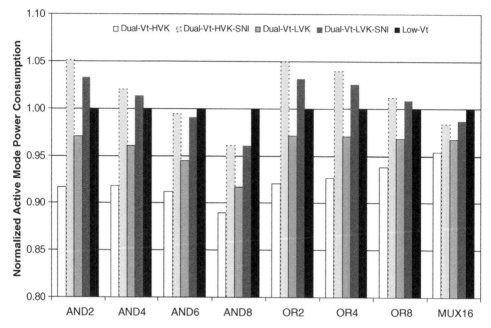

Figure 10.11   Comparison of the power consumption of domino logic circuits during the active mode. The power consumed by each gate is normalized to the power consumption of the corresponding low-V_t gate. SNI = similar noise immunity

10.4 DUAL THRESHOLD VOLTAGE CMOS TECHNOLOGY

The difference between the high and low threshold voltages (ΔV_t) is assumed in Sections 10.1, 10.2, and 10.3 to be 300 mV ($V_{tnlow} = |V_{tplow}| = 200$ mV and $V_{tnhigh} = |V_{tphigh}| = 500$ mV). The available high and low threshold voltages vary dramatically among different dual threshold voltage CMOS technologies [34], [35], [52], [142], [144]. The effect of the threshold voltages in a dual-V_t CMOS technology on the speed and power characteristics of the dual-V_t domino logic circuit technique is evaluated in this section. Under similar noise immunity conditions, the subthreshold leakage current, evaluation delay, precharge delay, and active mode power dissipation of the low-V_t and dual-V_t domino circuits are evaluated for three different sets of high and low threshold voltages (Tech$_1$: {$V_{tnlow} = |V_{tplow}| = 200$ mV and $V_{tnhigh} = |V_{tphigh}| = 300$ mV}, Tech$_2$: {$V_{tnlow} = |V_{tplow}| = 200$ mV and $V_{tnhigh} = |V_{tphigh}| = 400$ mV}, and Tech$_3$: {$V_{tnlow} = |V_{tplow}| = 200$ mV and $V_{tnhigh} = |V_{tphigh}| = 500$ mV).

For all three dual threshold voltage CMOS technologies, dual-V_t domino logic circuits based on an HVK transistor consume less subthreshold leakage energy as compared to dual-V_t domino logic circuits based on an LVK, under similar noise immunity conditions. In this section, therefore, a comparison of the electrical properties of only the dual-V_t domino logic circuits based on an HVK transistor and standard low-V_t domino logic circuits is presented.

The range of savings in subthreshold leakage current provided by the dual-V_t domino logic circuit technique for three different sets of dual threshold voltages is shown in Figure 10.12.

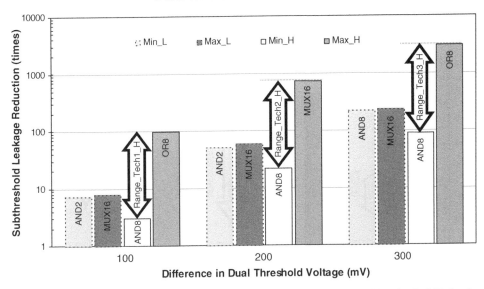

Figure 10.12   The range of savings in subthreshold leakage current provided by the dual-V_t domino logic circuit technique as compared to the standard low-V_t domino logic circuit technique for three different sets of dual threshold voltages. Min_L and Max_L = minimum and maximum, respectively, of the reduction in subthreshold leakage current as compared to the low-V_t domino logic circuits at a low dynamic node voltage state. Min_H and Max_H = minimum and maximum, respectively, of the reduction in subthreshold leakage current as compared to the low-V_t domino logic circuits at a high dynamic node voltage state

For Tech$_3$, the subthreshold leakage current of the dual-V_t domino logic circuits at a low dynamic node voltage is 224 times (AND8) to 235 times (MUX16) smaller as compared to the low-V_t domino logic circuits at a low dynamic node voltage state. For a smaller ΔV_t, the difference between $I_{Leak\text{-}HVPU}$, $I_{Leak\text{-}HVK}$, and $I_{Leak\text{-}HVN}$ and $I_{Leak\text{-}LVPU}$, $I_{Leak\text{-}LVK}$, and $I_{Leak\text{-}LVN}$, respectively, is also smaller. The achievable savings in subthreshold leakage energy is, therefore, reduced with the decreased difference in the dual threshold voltages. As illustrated in Figure 10.12, when the difference between the high and low threshold voltages is scaled to 100 mV (Tech$_1$), the subthreshold leakage current of the dual-V_t domino logic circuits is 7.3 times (AND2) to 7.9 times (MUX16) smaller as compared to the low-V_t domino logic circuits at a low dynamic node voltage.

   As compared to the low-V_t domino logic circuits at a high dynamic node voltage state, the subthreshold leakage current of the dual-V_t domino logic circuits at a low dynamic node voltage state is 89 times (AND8) to 3079 times (OR8) smaller for Tech$_3$. When ΔV_t is decreased to 100 mV (Tech$_1$), the subthreshold leakage current of the dual-V_t domino logic circuits is three times (AND8) to 99 times (OR8) smaller as compared to the low-V_t domino logic circuits with a high dynamic node voltage. For all three dual threshold voltage technologies, the effectiveness of the dual-V_t circuit technique for reducing subthreshold leakage current is greater in wide fan-in OR and AND–OR types of gates.

   The range of difference in the evaluation delay of the dual-V_t circuits as compared to the low-V_t circuits is shown in Figure 10.13. All of the delay differences are evaluated as a percentage of the delay of the corresponding low-V_t gate. A negative difference indicates a

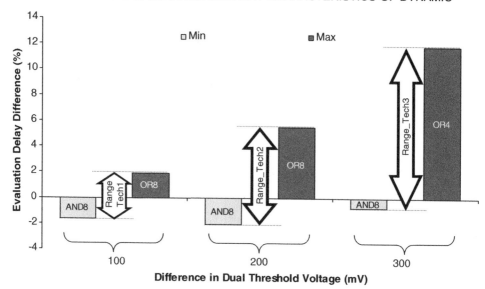

Figure 10.13 The range of difference in evaluation delay for the dual-V_t circuits as compared to the low-V_t domino logic circuits for three different sets of dual threshold voltages. A negative difference indicates a smaller evaluation delay as compared to a low-V_t circuit. Min = minimum difference in evaluation delay as compared to the low-V_t domino logic circuits. Max = maximum difference in evaluation delay as compared to the low-V_t domino logic circuits

higher evaluation speed as compared to a low-V_t circuit. For Tech$_3$, the difference between the evaluation delay of the dual-V_t and low-V_t domino logic circuits varies between -0.8% (AND8) and 11.8% (OR4). When ΔV_t is reduced to 100 mV, the difference between the evaluation delays varies between -1.7% (AND8) and 1.9% (OR8).

The range of difference in the precharge delay for dual-V_t circuits as compared to low-V_t circuits is shown in Figure 10.14. For Tech$_3$, the difference between the precharge delay of the dual-V_t and low-V_t domino logic circuits varies between 11.4% (AND8) and 23.3% (OR8). For Tech$_1$, the difference of the precharge delay varies between 5.7% (AND8) and 10.6% (OR2).

The range of difference between the power consumed by the dual-V_t and low-V_t circuits during the active mode of operation is shown in Figure 10.15. All of the power differences are evaluated as a percentage of the power consumption of the corresponding low-V_t gate. A negative difference indicates lower power consumption as compared to a low-V_t circuit. For Tech$_3$, the difference between the power consumed by the dual-V_t and low-V_t domino logic circuits varies between -3.9% (AND8) and 5.1% (AND2). For the six and eight input AND gates and the 16-bit multiplexer, the dual-V_t domino logic circuit technique reduces the power consumption during the active mode as well as the standby mode. When ΔV_t is reduced to 100 mV, the difference in the power consumption varies between -1.9% (AND8) and 0% (AND2). For Tech$_1$, the dual-V_t domino logic circuit technique reduces the power consumption during the active mode as well as the standby mode as compared to the low-V_t circuit technique for all of the domino gates (except for the two input AND gate for which the active power consumption of the low-V_t and dual-V_t circuit techniques is similar).

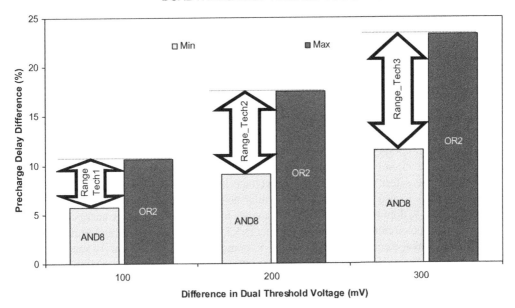

Figure 10.14 The range of difference in precharge delay between the dual-V_t and low-V_t domino logic circuits for three different sets of dual threshold voltages. Min = minimum difference in precharge delay as compared to low-V_t domino logic circuits. Max = maximum difference in precharge delay as compared to low-V_t domino logic circuits

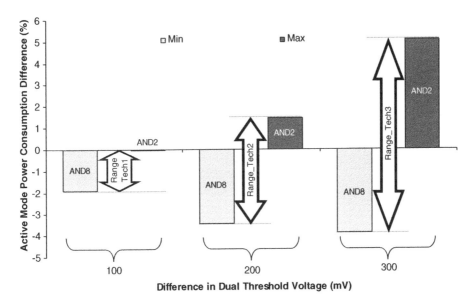

Figure 10.15 The range of difference in the power consumption (during the active mode) of the dual-V_t and low-V_t domino logic circuits for three different sets of dual threshold voltages. A negative difference indicates smaller power consumption as compared to a low-V_t circuit. Min = minimum difference in power consumption as compared to low-V_t domino logic circuits. Max = maximum difference in power consumption as compared to low-V_t domino logic circuits

10.5 SUMMARY

The node voltage-dependent subthreshold leakage current characteristics of domino logic circuits are examined in this chapter. A discharged dynamic node is preferred for reducing the leakage current in a dual-V_t domino logic circuit. Alternatively, a charged dynamic node is preferable for smaller subthreshold leakage energy in a standard low-V_t domino logic circuit with stacked pull-down devices, such as an AND gate.

Proper keeper and output inverter sizes are required in a dual-V_t domino logic circuit with a high-V_t keeper in order to maintain similar noise immunity as compared to a standard low-V_t domino logic circuit. As an alternative dual-V_t domino technique for enhanced noise immunity, the effect of a low threshold voltage keeper transistor on the leakage current characteristics is also evaluated. Under similar noise immunity conditions as compared to standard low-V_t domino logic circuits, the savings in subthreshold leakage energy offered by dual-V_t domino circuits with a high-V_t keeper is 5.7 to 10.9 times greater as compared to the savings in leakage current offered by dual-V_t domino circuits with a low-V_t keeper.

Under similar noise immunity conditions, the subthreshold leakage current of dual-V_t domino logic circuits with a low dynamic node voltage is 224 to 235 times smaller as compared to low-V_t domino logic circuits with a low dynamic node voltage. Alternatively, as compared to low-V_t domino logic circuits with a high dynamic node voltage, the subthreshold leakage current of dual-V_t domino logic circuits with a low dynamic node voltage is 89 to 3079 times smaller. The degradation in the precharge and evaluation speed of dual-V_t domino circuits is less than 23.4% and 11.8%, respectively, as compared to standard low-V_t domino circuits. The increase in active mode power consumption is less than 5.1%.

The effect of the difference of the high and low threshold voltages provided in a dual-V_t CMOS technology on the speed and power characteristics of the dual-V_t domino logic circuit technique is also evaluated. This technique provides significant savings in subthreshold leakage energy down to a 100 mV difference between the high and low threshold voltages in a 0.18 μm CMOS technology.

11 Sleep Switch Dual Threshold Voltage Domino Logic

The subthreshold leakage current characteristics of domino logic circuits are examined in Chapter 10. A dual threshold voltage domino logic circuit consumes significantly lower subthreshold leakage energy at a low dynamic node voltage state as compared to a high dynamic node voltage state. The dynamic node voltage-dependent asymmetry of the sub-threshold leakage current charcteristics of dual threshold voltage (dual-V_t) domino gates was first noted in [136]. A circuit technique to exploit this asymmetry has also been presented in [136]. Gating all of the inputs of the first stage of a domino pipeline is proposed in order to force dual-V_t domino gates into a low leakage sleep state [136].

The energy and delay overhead for entering and leaving the sleep mode, however, has not been addressed in [136]. Due to the additional gates at the inputs, significant dynamic switching energy is consumed to activate the sleep mode with the technique described in [136]. Additional energy is dissipated to precharge all of the dynamic nodes while reactivating a domino logic circuit at the end of an idle period. In order to justify the use of additional circuitry to place a dual-V_t circuit into a low leakage state, the total energy consumed to enter and leave the standby mode must be significantly less than the savings in standby leakage energy. Gating all of the inputs of the first stage of a domino circuit in a domino pipeline also increases the circuit area and active mode power. Furthermore, the circuit performance during the active mode is degraded due to the additional gates at the inputs. A circuit technique with low delay and energy overhead for placing a dual-V_t domino logic circuit into a low leakage state is, therefore, highly desirable.

A circuit technique is described in this chapter for lowering the subthreshold leakage energy consumption of domino logic circuits. The circuit technique employs sleep switches and a dual threshold voltage CMOS technology in order to place an idle domino logic circuit into a low leakage state. An 8-bit domino carry lookahead adder has been designed based on this circuit technique. The sleep switch circuit technique reduces the leakage energy by up to 830 times as compared to a standard low threshold voltage (low-V_t) domino circuit. The

Multi-Voltage CMOS Circuit Design V. Kursun and E. Friedman
© 2006 John Wiley & Sons, Ltd

dual threshold voltage domino adder enters and leaves the sleep mode within a single clock cycle. The sleep switch circuit technique enhances the effectiveness of a dual-V_t CMOS technology to reduce the subthreshold leakage current by strongly turning off all of the high threshold voltage (high-V_t) transistors, independent of the input vector. The sleep switch circuit technique lowers the subthreshold leakage energy by up to 714 times as compared to a standard dual-V_t domino logic circuit. The energy overhead of the circuit technique is low, permitting the sleep transistors to be activated during idle periods as short as 57 clock cycles so as to reduce the total power consumption.

Previously published leakage control techniques applicable to dual-V_t domino logic circuits are discussed in Section 11.1. The operation of the sleep switch dual-V_t domino logic circuit technique is described in Section 11.2. Simulation results characterizing the standby leakage energy and active mode delay and power of the sleep switch technique as compared to standard dual-V_t and low-V_t domino circuits are presented in Section 11.3. The noise immunity characteristics of the sleep switch dual-V_t domino logic circuit technique is evaluated in Section 11.4. The topics presented in this chapter are summarized in Section 11.5.

11.1 EXISTING SLEEP MODE CIRCUIT TECHNIQUES

Standard low-V_t and dual-V_t domino logic circuits are shown in Figure 11.1. As discussed in Chapter 10, if all of the high-V_t transistors are cut off in a dual-V_t domino logic circuit, the leakage current is significantly reduced as compared to a low-V_t circuit. The clock is gated high, cutting off the high-V_t pull-up transistors when a domino logic circuit is idle. In a standard dual-V_t domino logic circuit, the modes of operation of the high-V_t transistors other than the pull-up transistors are determined by the input vectors applied after the clock is gated high.

Subthreshold leakage current exponentially decreases with increasing threshold voltage. The leakage current of a cut-off high-V_t transistor is orders of magnitude lower as compared to a low-V_t transistor [29], [145]. Assuming a subthreshold slope of 85 mV/decade (a typical number in current CMOS technologies [52], [54]), an 85 mV increase in the threshold voltage of a transistor reduces the subthreshold leakage current by ten times. Leakage currents in a dual-V_t circuit can be reduced by employing a greater number of high-V_t transistors [145]. Unless all of the high-V_t transistors are strongly cut off, the potential savings in energy from the dual-V_t domino circuit technique cannot be fully exploited. Circuit techniques to place a domino logic circuit into a low leakage state regardless of the input vectors and the initial voltage state of the circuit nodes (before the clock is gated) are desirable. Dual-V_t domino logic circuit techniques with different standby control mechanisms have been proposed in the literature [34], [35], [130], [136], [141], [142].

A dual-V_t circuit technique was proposed in [136] to reduce leakage current in domino pipelines. The technique described in [136] requires the input signals of the first stage circuits in a domino pipeline to be gated. After forcing the first stage of the domino gates to evaluate and discharge, the domino gates of the subsequent stages in the pipeline also evaluate and discharge in a domino fashion. The technique proposed in [136], however, is ineffective in placing a circuit into a low leakage state if some of the domino gates in a cascaded domino logic circuit require inverted signals (such as an XOR domino gate generating a sum bit at the output stage of a domino carry lookahead adder). Most domino logic circuits cannot be placed

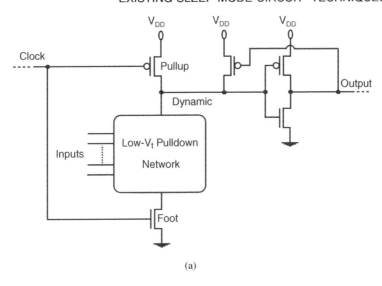

(a)

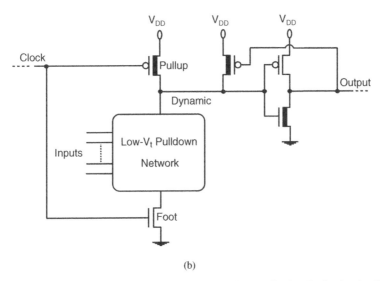

(b)

Figure 11.1 Standard domino logic circuits. (a) Standard low-V_t domino logic circuit. (b) Standard dual-V_t domino logic circuit. High-V_t transistors are symbolically represented by a thick line in the channel region

into a minimum leakage state in which all of the high-V_t transistors are strongly cut off simply by gating the input vectors of the first stage of a domino circuit. The technique proposed in [136] also requires a significant dynamic switching energy overhead for activating the sleep mode due to the additional gates at the inputs. The dual-V_t domino circuit proposed in [136] only offers savings in energy if the circuit stays idle for a long time. Furthermore, gating all of the inputs of the first stage of a domino circuit in a domino pipeline increases the circuit area and active mode power. The circuit performance during the active mode is also degraded due to the additional gates at the inputs of the domino circuit.

An alternative dual-V_t technique was proposed in [130] to reduce the dynamic power, propagation delay, and area overhead as compared to the technique proposed in [136]. Although the delay and area overhead is reduced by the technique proposed in [130], the energy consumed during the standby mode is higher as compared to the circuit proposed in [136]. This increased standby energy is primarily due to the NMOS transistor inside the output inverter of the domino gates in the first stage of each domino pipeline not being completely turned off and because the keeper has a low-V_t in the technique described in [130]. Dual-V_t domino logic circuits based on a low-V_t keeper transistor consume significantly higher subthreshold leakage energy as compared to dual-V_t circuit techniques based on a high-V_t keeper transistor (see Chapter 10 for a quantitative comparison of the two circuit techniques).

An approach which utilizes the leakage currents of the pull-down path transistors has been proposed in [142] in order to place a dual-V_t domino logic circuit into a low leakage state. High-V_t switches are employed in series with the keeper and the NMOS transistor of the output inverter in a domino circuit. When the circuit is active, these high-V_t switches are on and the circuit operates similarly to a standard dual-V_t circuit. When the circuit is idle, the high-V_t series switches are cut off by a sleep signal, isolating the dynamic node from the power supply. The floating dynamic node slowly discharges due to the leakage current of the transistors along the pull-down path. The high-V_t switch in series with the NMOS transistor of the output inverter ensures that no short-circuit power is consumed during the slow discharge of the dynamic node. A high-V_t series transistor at the output inverter, however, degrades the precharge delay. Furthermore, a high-V_t transistor in series with a keeper degrades the noise immunity. To minimize the degradation in noise immunity and precharge delay, the size of these series switches should be increased. Wider series transistors, however, increase the energy overhead of activating the standby leakage control mechanism. Increasing the series transistor size also increases the area overhead of this technique. Another disadvantage of this technique is the low speed of the proposed mechanism for placing a circuit into a low leakage state. The circuit technique proposed in [142], therefore, may not be feasible for fine-grain leakage reduction during short idle periods (a few tens to hundreds of clock cycles) in high performance integrated circuits.

11.2 DUAL THRESHOLD VOLTAGE DOMINO LOGIC EMPLOYING SLEEP SWITCHES

A low energy and delay overhead circuit technique is presented in this chapter to lower the subthreshold leakage currents in an idle domino logic circuit. The circuit technique employs sleep switches to place a dual-V_t domino logic circuit into a low leakage state within a single clock cycle. A domino logic circuit based on the sleep switch dual-V_t circuit technique is shown in Figure 11.2.

A high-V_t NMOS switch is added to the dynamic node of a domino circuit as shown in Figure 11.2. The operation of this transistor is controlled by a separate sleep signal. During the active mode of operation, the sleep signal is set low, the sleep switch is cut off, and the dual-V_t circuit operates as a standard dual-V_t domino circuit. During the standby mode of operation, the clock signal is maintained high, turning off the high-V_t pull-up transistor of each domino gate. The sleep signal transitions high, turning on the sleep switch. The

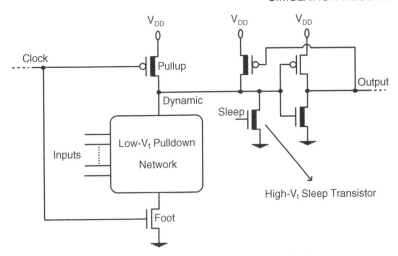

Figure 11.2 Sleep switch dual-V_t domino logic circuit technique. High-V_t transistors are symbolically represented by a thick line in the channel region

dynamic node of the domino gate is discharged through the sleep switch, thereby turning off the high-V_t NMOS transistor within the output inverter. The output transitions high, cutting off the high-V_t keeper. Following the low-to-high transition at the output of a sleep switch dual-V_t domino gate, the subsequent gates (fed by the non-inverting signals) also evaluate and discharge in a domino fashion. After the node voltages settle to a steady state, all of the high-V_t transistors are strongly cut off, significantly reducing the subthreshold leakage current. Note that this technique, requiring no additional gating on the input signals while strongly turning off all of the high-V_t transistors within a single clock cycle, is significantly more power, delay, and area efficient as compared to the techniques proposed in [130], [136], and [142].

11.3 SIMULATION RESULTS

Eight input clock-delayed domino carry lookahead adders based on the low-V_t, standard dual-V_t, and sleep switch circuit techniques are evaluated assuming a 0.18 µm CMOS technology ($V_{tnlow} = |V_{tplow}| = 200$ mV, $V_{tnhigh} = |V_{tphigh}| = 500$ mV, and $T = 110°$C). The block diagram of a clock-delayed domino carry lookahead adder based on the sleep switch dual-V_t circuit technique is shown in Figure 11.3. Each sum output drives a capacitive load of 10 fF. A 1 GHz clock with a 50% duty cycle is applied to the domino logic circuits. All of the common transistors in the sleep switch and standard dual-V_t adders are sized the same.

In the sleep switch adder, all of the propagate (P), generate (G), and sum (S) domino gates have sleep switches. When the domino adder is idle, the dynamic nodes of the P and G domino gates (in the first stage propagate and generate (PG) block) are forced to discharge via sleep switches. The domino gates within the lookahead carry (C) block do not have sleep switches. Following the low-to-high transition at the output of the P and G gates, the domino

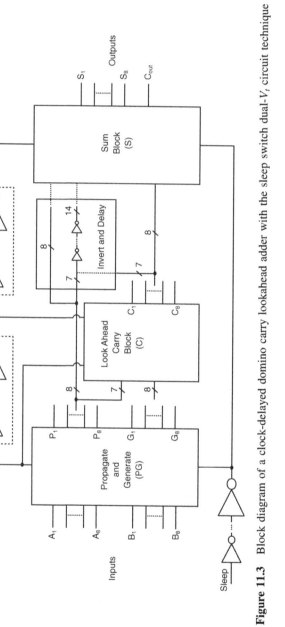

Figure 11.3 Block diagram of a clock-delayed domino carry lookahead adder with the sleep switch dual-V_t circuit technique

Table 11.1   Input Vectors Applied to an Adder

	V_0	V_1	V_2	V_3	V_4	V_5
A	0	0	1	1	255	255
B	0	255	255	127	255	0

gates within the carry block also evaluate and discharge in a domino fashion. Some of the signals originating from the PG and C blocks are inverted before being fed into the sum block (see Figure 11.3). The domino logic circuits within the sum block, therefore, also require sleep switches in order to place the circuits into a low leakage state.

The input vectors applied to the adder are listed in Table 11.1. The leakage characteristics of the circuits are evaluated for six input vectors, V_0 to V_5. C_{out} (S_8) is evaluated through the critical path of the carry chain within the carry block for the input vector V_2 (V_3). The delay and active mode power are calculated for V_2 and V_3.

The sleep switch circuit technique significantly reduces the subthreshold leakage current as compared to both low-V_t and standard dual-V_t circuits. The standby leakage energy characteristics of the adders based on the standard dual-V_t, low-V_t, and sleep switch circuit techniques are presented in Section 11.3.1. The subthreshold leakage current characteristics of a standard domino logic circuit display a strong dependence on the input vectors when the circuit is at a high dynamic node voltage state. The stack effect on the subthreshold leakage current characteristics of a domino logic circuit at a high dynamic node voltage state is described in Section 11.3.2. The sleep switch circuit technique also enhances the active mode delay and power characteristics as compared to a low-V_t circuit. The active mode delay and power characteristics of the circuit technique are discussed in Section 11.3.3. The sleep/wake-up delay and energy overhead are presented in Section 11.3.4.

11.3.1 Subthreshold Leakage Energy Reduction

The standby leakage energy characteristics of the low-V_t, standard dual-V_t, and sleep switch dual-V_t adders are evaluated in this section. When a low-V_t, standard dual-V_t, or sleep switch domino logic circuit is idle, the clock is gated high. In a sleep switch circuit, the sleep transistors are activated after clock gating. The leakage energy consumption (per clock cycle) of the low-V_t, standard dual-V_t, and sleep switch adders is shown in Figure 11.4.

The leakage energy of a standard dual-V_t circuit is reduced by 1.2 to 2.8 times as compared to a low-V_t circuit. The standby leakage energy of the standard low-V_t and dual-V_t circuits is dependent on the applied input vector after the clock signal is gated high. The dynamic nodes of all of the domino logic circuits are precharged when the clock is low. After the clock transitions high, several of these domino gates evaluate and discharge, provided that a necessary input combination to discharge the dynamic node is applied. A high dynamic node voltage state is typically the highest subthreshold leakage state for a dual-V_t domino logic gate since all of the high-V_t transistors (other than the pull-up transistors) operate in the strong inversion region. As discussed in Section 11.1, the advantages of a dual-V_t CMOS technology for reducing leakage current are maximized when all of the high-V_t transistors are strongly cut off during the idle mode. For V_0, the dynamic nodes of all of the domino logic gates of a standard dual-V_t adder are maintained high during the idle mode. The V_0

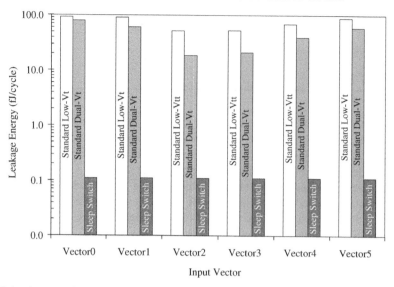

Figure 11.4   A comparison of the leakage energy (per clock cycle) of the adder circuits with the low-V_t, standard dual-V_t, and sleep switch circuit techniques for six different input vectors

vector, therefore, produces the maximum subthreshold leakage current in a standard dual-V_t adder. For V_0, the subthreshold leakage current is produced by the low-V_t transistors rather than the high-V_t transistors as in a standard low-V_t adder. The subthreshold leakage current of the domino gates within the standard low-V_t and dual-V_t adders is, therefore, similar for V_0. The small difference between the subthreshold leakage current characteristics of the standard low-V_t and dual-V_t adders for V_0 is due to the reduced leakage of the dual-V_t delay elements in a standard dual-V_t domino adder.

As shown in Figure 11.4, the sleep switch circuit technique minimizes the leakage energy for all of the input vectors as compared to both the low-V_t and standard dual-V_t circuits. By activating the sleep transistors, all of the domino gates are placed into a low leakage state for any given input vector. The reduction in leakage energy offered by the sleep switch circuit technique varies between 461 times (V_2) and 830 times (V_0) as compared to a low-V_t adder. The circuit technique enhances the effectiveness of a dual-V_t CMOS technology in reducing subthreshold leakage current by cutting off all of the high-V_t transistors. An adder based on the sleep switch circuit technique dissipates 167 times (V_2) to 714 times (V_0) lower leakage energy as compared to a standard dual-V_t adder.

11.3.2 Stack Effect in Domino Logic Circuits

For V_1 and V_5, the dynamic nodes of the generate and carry gates are maintained high while the propagate and sum gates evaluate and discharge in a standard domino adder. In a standard dual-V_t adder, the subthreshold leakage current in the propagate and sum gates is 2415 and 1149 times, respectively, smaller for both V_1 and V_5 as compared to V_0. Similarly, in a standard low-V_t adder, the subthreshold leakage current in the propagate and sum gates is 3.3 and 1.7 times, respectively, smaller for V_1 and V_5 as compared to V_0. Despite this significant reduction in the subthreshold leakage current of the propagate and sum gates, the second and third highest leakage currents in standard low-V_t and dual-V_t domino logic circuits are observed for V_1 and V_5, respectively, as shown in Figure 11.4. The subthreshold leakage current of the generate and carry gates in standard dual-V_t and low-V_t adders approximately doubles for V_1 and V_5 as compared to V_0. This significant increase in subthreshold leakage current for a high dynamic node voltage state of the generate and carry gates is caused by the stack effect [143], [146].

Input vectors applied after clock gating determine which transistors produce subthreshold leakage current together with the voltage state of the dynamic node. As discussed previously, a high dynamic node voltage state is typically the highest leakage state in a dual-V_t domino logic gate. A variation in the sources of the subthreshold leakage current in the pull-down network of a standard dual-V_t domino (generate) gate with the input vector for a high voltage state of the dynamic node is shown in Figure 11.5. The dynamic node voltage in the generate and carry gates is maintained high for three different input vectors, V_0, V_1, and V_5, as shown in Figures 11.5(a), (b), and (c), respectively.

For V_0, both N_1 and N_2 operate in the weak inversion region. The voltage at node$_1$ rises until the subthreshold leakage currents through N_1 and N_2 are equal (in the steady-state condition). The total subthreshold leakage current at steady state is

$$I_{subthreshold-H} = I_{Leak-PD} + I_{Leak-P1}, \tag{11.1}$$

$$I_{Leak-PD} = I_{Leak-N1-V0} = I_{Leak-N2-V0}, \tag{11.2}$$

where $I_{Leak-P1}$ is the subthreshold leakage current through the low-V_t pull-up transistor within the output inverter. $I_{Leak-N1-V0}$ and $I_{Leak-N2-V0}$ are the subthreshold leakage currents through N_1 and N_2, respectively, for V_0.

For V_1, N_1 operates in the weak inversion region. Alternatively, N_2 operates in the strong inversion region. The total subthreshold leakage current at steady state is

$$I_{subthreshold-H} = I_{Leak-N1-V1} + I_{Leak-P1}, \tag{11.3}$$

where $I_{Leak-N1-V1}$ is the subthreshold leakage current through N_1 for V_1.

For V_5, N_2 operates in the weak inversion region while N_1 is turned on (strong inversion). The total subthreshold leakage current at steady state is

$$I_{subthreshold-H} = I_{Leak-N2-V5} + I_{Leak-P1}, \tag{11.4}$$

where $I_{Leak-N2-V5}$ is the subthreshold leakage current through N_2 for V_5.

For V_0, N_1 and N_2 are cut off. A steady-state voltage is reached when the voltage at node$_1$ rises to approximately 41 mV above ground, equalizing the subthreshold leakage currents through N_1 and N_2. The subthreshold leakage current through a stack of cut-off transistors is significantly smaller than the subthreshold leakage current through a single cut-off transistor [143], [146]. The subthreshold leakage current of a MOSFET is exponentially dependent on the threshold, gate-to-source, and drain-to-source voltages [39], [48]. The subthreshold leakage current through N_1 exponentially decreases for V_0 as compared to V_1, due to increased threshold voltage (reverse body bias), negative gate-to-source voltage, and lower drain-to-source voltage. For V_1, the voltage at node$_1$ and node$_2$ are both at approximately ground level since N_2 operates in the strong inversion region. This condition eliminates the reverse

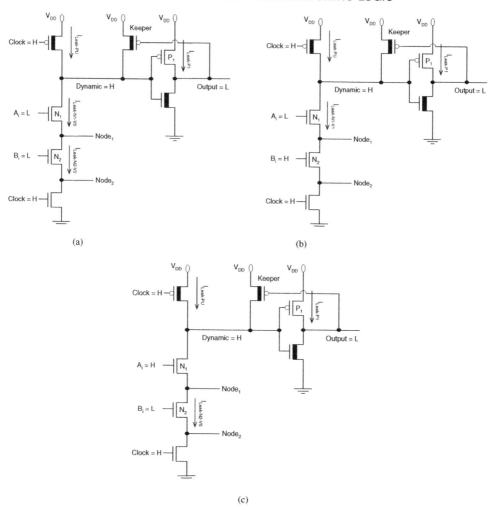

Figure 11.5   Variation of subthreshold leakage current conduction paths with input vector for a high voltage state at the dynamic node in a standard dual-V_t domino logic circuit. (a) Sources of subthreshold leakage current for V_0. (b) Sources of subthreshold leakage current for V_1. (c) Sources of subthreshold leakage current for V_5. H: high. L: low

body bias and negative gate-to-source voltage while increasing the drain-to-source voltage of N_1. $I_{Leak\text{-}N1\text{-}V1}$ is, therefore, higher than $I_{Leak\text{-}N1\text{-}V0}$. The total subthreshold leakage current of the generate gates is 2.3 times higher for V_1 than V_0. Similarly, the subthreshold leakage current of the carry gates is 1.7 times higher for V_1 than V_0. For V_5, N_1 is on while N_2 is cut off. The voltage at node$_1$ is high, increasing the drain-to-source voltage of N_2. $I_{Leak\text{-}N2\text{-}V5}$ is, therefore, higher than $I_{Leak\text{-}N2\text{-}V0}$, increasing the total subthreshold leakage current of the generate gates by 2.1 times. Similarly, the subthreshold leakage current of the carry gates is 1.7 times greater for V_5 than V_0. The subthreshold leakage current in the standard low-V_t carry and generate gates also significantly decrease for V_0 as compared to V_1 and V_5, due to the stack effect.

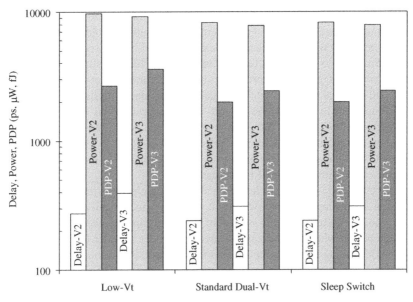

Figure 11.6 A comparison of the delay, power, and PDP of adder circuits with low-V_t, standard dual-V_t, and sleep switch circuit techniques for the input vectors V_2 and V_3

11.3.3 Delay and Power Reduction in the Active Mode

The active mode delay, power, and power–delay product (PDP) of low-V_t, standard dual-V_t, and sleep switch adders are shown in Figure 11.6. The delay and power characteristics of a standard dual-V_t adder are similar to the sleep switch adder. The sleep switch circuit technique enhances the evaluation speed by 12% and 21%, for V_2 and V_3, respectively, as compared to a low-V_t adder. The enhancement in speed with the sleep switch circuit technique is primarily due to the reduced contention current [33] of a high-V_t keeper (see Chapter 9 for a discussion of contention current).

The sleep switch circuit technique also reduces the active mode power consumption as compared to a low-V_t circuit. The power consumption is reduced by 14.4% and 14.6% for the input vectors V_2 and V_3, respectively, as compared to a low-V_t adder. A portion of the savings in active mode power is due to the reduced contention current of the high-V_t keeper transistor in a dual-V_t circuit (see Figures 11.1 and 11.2). Another important factor that reduces the power consumption of a dual-V_t circuit is the lower power consumption in dual-V_t delay elements.

11.3.4 Sleep/Wake-Up Delay and Energy Overhead

When a sleep switch domino logic circuit is idle, the clock is gated high. The sleep signal is applied to a domino stage after the low-to-high edge of the clock signal propagates to the corresponding stage of a clock-delayed domino logic circuit. Activating the sleep switches after the arrival of the low-to-high transition of the clock signal ensures that no short-circuit power is consumed while entering the sleep mode. Activating the sleep switches forces all of the domino gates into a low dynamic node voltage state. After the node voltages settle, all of the high-V_t transistors are strongly cut off, decreasing the subthreshold leakage currents with the sleep switch circuit technique. Less than a clock period is required (depending upon the input vector, from 829 ps to 850 ps after the clock is gated) for the entire adder circuitry to be placed into a low leakage state. Before the end of an idle mode, the sleep signal transitions low, cutting off all of the sleep switches. Disabling the sleep transistors before activating the clock is important in order to avoid short-circuit currents when leaving the idle mode. The clock is reactivated and all of the dynamic nodes are recharged to activate (wake up) a sleeping domino circuit. The duration of reactivation is equal to the precharge time of a domino circuit. An adder circuit, therefore, is able to enter and leave the standby mode within a single clock cycle with this circuit technique.

The energy overhead to enter and leave the sleep mode with the sleep switch technique is also evaluated. Activating the sleep switches in order to place a dual-V_t domino logic circuit into standby mode requires a specific amount of energy. Additional energy is dissipated at the end of an idle period while precharging the dynamic nodes in order to reactivate a domino logic circuit. Depending upon the input vectors, some or none of the dynamic nodes in the low-V_t and standard dual-V_t circuits are discharged during the sleep mode. Alternatively, all of the dynamic nodes in a sleep switch domino logic circuit are discharged during the sleep mode, independent of the input vectors. The activation energy required by the sleep switch circuit technique is, therefore, higher than the low-V_t and standard dual-V_t circuit techniques. In order to justify the sleep switch circuit technique to force a dual-V_t circuit into a low leakage state, the total energy required to enter and leave the sleep mode must be less than the total savings in standby leakage energy.

The cumulative energy dissipated in the standby mode by the low-V_t and sleep switch adders is shown in Figure 11.7. It is assumed that the junction temperature does not

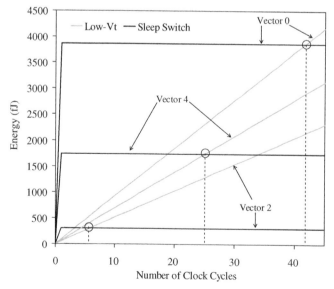

Figure 11.7 Cumulative standby energy dissipation of the low-V_t and sleep switch adders for three different input vectors

significantly change for the duration of the standby mode. The leakage energy per cycle is also assumed to be constant. The cumulative energy of a low-V_t domino circuit is only affected by subthreshold leakage current during the standby mode. Alternatively, both the cumulative leakage energy and the energy overhead of entering and leaving the sleep mode are included in the energy characteristics of the sleep switch circuit. The total energy overhead of the sleep switch circuit technique is independent of the duration of the idle mode. The energy overhead for employing the sleep switch circuit technique is dissipated even if the domino circuit remains in the standby mode for only a single clock cycle. The total energy overhead of the technique (composed of the energy dissipated in order to activate the sleep transistors while entering the sleep mode and disable the sleep transistors and reactivate the domino gates after the standby mode is over) is included as an energy step in the first cycle of the standby mode (see Figure 11.7). Similar to the low-V_t energy characteristics, after the first clock cycle, the sleep switch circuit energy is only due to the subthreshold leakage current. Since the standby leakage energy of a sleep switch circuit is significantly lower (up to 830 times) than a low-V_t circuit, the sleep switch energy characteristics have a much smaller slope as compared to the energy characteristics of the low-V_t adder (see Figure 11.7). A specific amount of time in the idle mode, also dependent upon the input vectors, is necessary for the cumulative leakage energy of a low-V_t circuit to exceed the cumulative energy of a sleep switch circuit.

The intersection of the sleep switch and low-V_t cumulative energy characteristics are evaluated to determine the necessary minimum duration of the sleep mode of operation such that the sleep switch circuit technique offers a net saving in energy as compared to a low-V_t circuit. As shown in Figure 11.7, the cumulative standby energy of the low-V_t and sleep switch circuits exhibits different behavior depending upon the input vectors. The leakage current of a low-V_t adder is smallest for V_2 and highest for V_0 (see Figure 11.4). Alternatively, the leakage current of a sleep switch adder is virtually independent of the input vectors. Depending upon the input vectors, the energy overhead of the sleep switch scheme changes. For V_0, none of the dynamic nodes of a low-V_t circuit are discharged during the standby mode. Alternatively, all of the dynamic nodes are discharged in a sleep switch circuit. The relative energy overhead of the sleep switch circuit technique required to charge the dynamic nodes to reactivate the circuit (to transition from standby mode to active mode) is, therefore, highest for V_0. As shown in Figure 11.7, a minimum of 42 clock cycles is required for the sleep switch circuit technique to provide a net saving in energy as compared to a low-V_t circuit during standby mode.

As discussed previously, a standard dual-V_t circuit offers savings in leakage current of 1.2 to 2.8 times as compared to a low-V_t circuit. The energy savings of a standard dual-V_t domino circuit originates from the selective replacement of a group of high leakage, low-V_t transistors with a group of low leakage, high-V_t transistors. Unlike the sleep switch circuit technique, a standard dual-V_t circuit does not introduce any energy overhead in order to reduce standby leakage current. Although the leakage energy of a sleep switch circuit is significantly reduced as compared to a standard dual-V_t circuit, the non-negligible energy overhead of the sleep switch circuit technique must also be assessed to accurately compare the energy characteristics of the two circuit techniques. The cumulative energy dissipated during standby mode by the sleep switch and standard dual-V_t adders is shown in Figure 11.8.

The (step) change in energy of the sleep switch characteristics during the first cycle represents the energy overhead for activating the sleep switches (to enter the sleep mode) and for deactivating the sleep switches and recharging the domino gates (to exit the sleep mode).

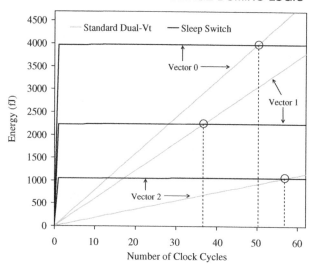

Figure 11.8   Cumulative standby energy dissipation of the sleep switch and standard dual-V_t adders for three different input vectors

Since the sleep switch circuit technique reduces the standby leakage energy by 167 to 714 times as compared to a standard dual-V_t circuit, the sleep switch characteristics have a significantly smaller slope after the first cycle as compared to the energy characteristics of a standard dual-V_t adder. As discussed previously, V_0 produces the highest leakage state in a standard dual-V_t circuit. Alternatively, the leakage energy of a standard dual-V_t adder is lowest for V_2. No input combination exists that can place a standard dual-V_t adder into a lower leakage state as compared to a sleep switch dual-V_t adder. Circuit techniques based on applying a selected input vector to place a circuit into a low leakage state (such as the technique described in [86] and [136]) are, therefore, ineffective in minimizing the leakage of the domino adder as discussed in this section.

As shown in Figure 11.8, a minimum of 57 clock cycles is required for the sleep switch circuit technique to provide a net saving in energy as compared to a standard dual-V_t circuit during the standby mode. Although the leakage energy of the standard dual-V_t domino adder is 167 to 714 times higher as compared to the sleep switch adder, a standard dual-V_t circuit technique is preferable in those applications with idle periods shorter than 57 clock cycles.

11.4   NOISE IMMUNITY COMPENSATION

As discussed in Chapter 9, in a standard domino logic gate, a feedback keeper is employed to maintain the state of the dynamic node against coupling noise, charge sharing, and subthreshold leakage current. In a dual-V_t domino logic circuit, the keeper transistor has a high-V_t (see Figures 11.1 and 11.2). As discussed in Chapter 10, the current supplied by a high-V_t keeper to preserve the state of a dynamic node is reduced, thereby degrading the noise immunity as compared to a low-V_t circuit. The degradation in noise immunity varies for different blocks within an adder.

During evaluation of the noise immunity characteristics, the same noise signal is coupled to each of the inputs of a domino logic circuit as this situation represents the worst case

Table 11.2 Degradation in Noise Immunity of Standard Dual-V_t and Sleep Switch Adders as Compared to the Low-V_t Adder with Same Size Transistors

	Gate	Propagate	Generate	Carry	Sum
Average reduction in	Standard dual-V_t	12.2%	14.0%	12.9%	11.3%
noise immunity	sleep switch	12.2%	14.9%	12.9%	11.3%

noise condition. In sleep switch circuits, the noise is also assumed to couple to the gates of the sleep transistors. The noise margin criterion used in this section is similar to the criterion described in [129]. The noise immunity is the voltage amplitude of the DC noise signal that produces a signal with the same amplitude at the output of a domino logic circuit, assuming a 1 GHz clock with a 50% duty cycle. The average degradation in noise immunity for the propagate, generate, carry, and sum domino logic gates is listed in Table 11.2. The degradation in noise immunity of the sleep switch domino logic gates varies between 11.3% and 14.9% as compared to the low-V_t circuits. The effect of the sleep switches on the noise immunity characteristics of the dual-V_t domino logic gates is small. Sleep switch propagate and sum gates have similar noise immunity as compared to the standard dual-V_t propagate and sum circuits. The noise immunity degradation in the sleep switch generate gates as compared to the standard dual-V_t generate gates is less than 1.7%.

Both keeper and output inverter sizing is required in a dual-V_t domino logic circuit with a high-V_t keeper transistor in order to provide similar noise immunity as compared to a standard low-V_t domino logic circuit (see Chapter 10 for a detailed discussion). An alternative technique for enhanced noise immunity is to employ a low-V_t keeper transistor in a dual-V_t domino circuit. Unless the output inverter is resized, a dual-V_t domino circuit with a low-V_t keeper transistor is not capable of providing noise immunity comparable to a low-V_t domino logic circuit. Under similar noise immunity conditions as compared to the standard low-V_t domino logic circuits, the savings in subthreshold leakage energy offered by a dual-V_t circuit technique based on a high-V_t keeper is significantly higher as compared to the savings in leakage current offered by a dual-V_t circuit technique based on a low-V_t keeper. In this section, therefore, the high-V_t keeper and output inverter pull-down transistor widths of each sleep switch and standard dual-V_t domino gate are increased so as to maintain a similar noise immunity as compared to standard low-V_t gates. A comparison of the subthreshold leakage energy (per clock cycle) of the low-V_t, standard dual-V_t, and sleep switch dual-V_t domino adders under similar noise immunity conditions for different input vectors is shown in Figure 11.9. The normalized leakage energy consumption of the low-V_t, standard dual-V_t, and sleep switch adders under similar and degraded noise immunity conditions is listed in Table 11.3.

Table 11.3 A Comparison of Normalized Subthreshold Leakage Energy of Low-V_t, Standard Dual-V_t, and Sleep Switch Adders Under Similar and Degraded Noise Immunity Conditions

		V_0	V_1	V_2	V_3	V_4	V_5
Similar transistor size–degraded	Standard low-V_t	830	802	461	473	622	790
noise immunity	Standard dual-V_t	714	546	167	187	355	535
	sleep switch	1	1	1	1	1	1
Transistor sizing for similar	Standard low-V_t	660	637	366	376	495	628
noise immunity	Standard dual-V_t	565	434	132	148	281	425
	sleep switch	1	1	1	1	1	1

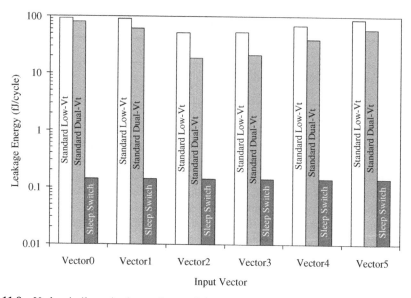

Figure 11.9   Under similar noise immunity conditions, a comparison of the leakage energy (per clock cycle) of the adder circuits with the low-V_t, standard dual-V_t, and sleep switch circuit techniques for six different input vectors

The subthreshold leakage energy consumed by a standard dual-V_t domino adder is determined by the subthreshold leakage current of the domino gates which are at a high dynamic node voltage state. When the dynamic node voltage is high, the subthreshold leakage current characteristics are virtually independent of the width of the keeper and output inverter pull-down transistors (see Chapter 10). Keeper and output inverter sizing for enhanced noise immunity, therefore, has little effect on the subthreshold leakage energy consumed by a standard dual-V_t adder, noticeable in the comparison illustrated in Figures 11.4 and 11.9.

The dynamic nodes of all of the domino gates in a sleep switch circuit are maintained in a low voltage state. In a dual-V_t domino logic circuit operating at a low dynamic node voltage state, the subthreshold leakage current strongly depends on the width of the keeper and output inverter pull-down transistors (see Chapter 10). The subthreshold leakage current in a sleep switch circuit, therefore, increases after keeper and output inverter sizing, degrading the savings in subthreshold leakage energy, as listed in Table 11.3. The subthreshold leakage current in the sleep switch adder is 366 to 660 times smaller as compared to the low-V_t adder under similar noise immunity conditions. The subthreshold leakage energy dissipation of the sleep switch dual-V_t adder is 132 to 565 times smaller as compared to the standard dual-V_t adder with similar noise immunity characteristics.

The cumulative energy dissipated in the standby mode by the low-V_t and sleep switch adders under similar noise immunity conditions is shown in Figure 11.10. The cumulative energy dissipated in the standby mode by the standard dual-V_t and sleep switch dual-V_t adders providing similar noise immunity characteristics is shown in Figure 11.11. The minimum duration of the idle mode required to provide a net saving in the total energy dissipation increases due to the higher subthreshold leakage currents in the sleep switch domino adder after transistor sizing. As shown in Figure 11.10, a minimum of 47 clock cycles is required for the sleep switch circuit to provide a net power savings as compared to a standard

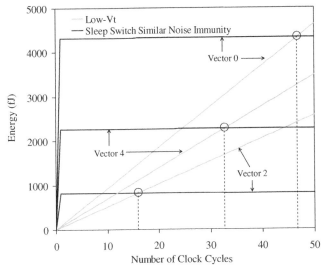

Figure 11.10   Under similar noise immunity conditions, cumulative standby energy dissipation of the low-V_t and sleep switch adders for three different input vectors

low-V_t adder during the idle mode. Similarly, as shown in Figure 11.11, a minimum of 69 clock cycles is required to provide a net saving in total standby energy consumption as compared to a standard dual-V_t domino adder. The minimum number of clock cycles required for the sleep switch circuit to provide a net savings in total energy during the idle mode, for both similar transistor sizing (degraded noise immunity) and similar noise immunity conditions (increased keeper and output inverter pull-down transistor size), is listed in Table 11.4.

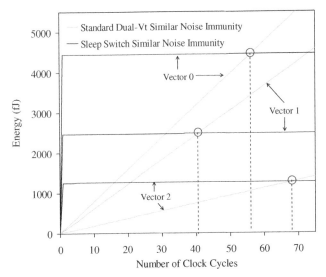

Figure 11.11   Under similar noise immunity conditions, cumulative standby energy dissipation of the sleep switch and standard dual-V_t adders for three different input vectors

Table 11.4 Minimum Duration of the Idle Mode Required for the Sleep Switch Circuit Technique to Provide a Net Saving in Standby Energy as Compared to the Standard Low-V_t and Dual-V_t Adders Under Similar and Degraded Noise Immunity Conditions

	Minimum Number of Clock Cycles Required			
	Similar Transistor Size–Degraded Noise Immunity		Transistor Sizing for Similar Noise Immunity	
Vector	Standard Low-V_t	Standard Dual-V_t	Standard Low-V_t	Standard Dual-V_t
V_0	42	51	47	57
V_1	20	37	25	41
V_2	6	57	16	69
V_3	9	57	19	68
V_4	26	56	33	61
V_5	23	42	29	47

11.5 SUMMARY

A circuit technique is presented in this chapter for reducing the standby leakage energy consumption of domino logic circuits. This circuit technique employs sleep switches and a dual threshold voltage CMOS technology in order to place an idle domino logic circuit into a low subthreshold leakage state, without degrading the delay and power characteristics during the active mode. A dual threshold voltage domino circuit enters and leaves the sleep mode within a single clock cycle with the sleep switch circuit technique.

The sleep switch circuit technique reduces the leakage energy by up to 830 times as compared to a standard low-V_t circuit. The circuit technique also reduces the active mode delay and power by up to 21% and 14.6%, respectively, as compared to a low-V_t circuit.

Existing techniques based on the application of a selected input vector to place a dual-V_t circuit into a low leakage state are ineffective for minimizing subthreshold leakage currents in multiple stage domino circuits with inverted internal signals. The sleep switch circuit technique unleashes the full potential of a dual-V_t CMOS technology to reduce subthreshold leakage current by strongly turning off all of the high-V_t transistors, independent of the input signals. The sleep switch circuit technique reduces leakage energy by up to 714 times as compared to a standard dual-V_t circuit. The energy overhead of the circuit technique is low, justifying the use of the sleep scheme during idle periods as short as 57 clock cycles in order to reduce standby leakage energy.

The noise immunity of the circuit blocks within a dual-V_t domino adder is degraded by up to 14.9% as compared to a low-V_t domino adder. The keepers and output inverter pull-down transistors are sized to provide similar noise immunity as compared to a low-V_t adder. Under similar noise immunity conditions, the subthreshold leakage energy consumed by a sleep switch adder is up to 660 times and 565 times smaller as compared to a standard low-V_t and dual-V_t adder, respectively. A minimum of 47 and 69 clock cycles is required for the sleep switch circuit to provide a net savings in total energy consumption during the idle mode while providing similar noise immunity characteristics during the active mode as compared to a standard low-V_t and dual-V_t domino adder, respectively.

12 Conclusions

The fundamental enabling force behind the evolution of integrated circuits is the advancements in semiconductor fabrication technologies that simultaneously permit scaling the minimum feature size of the device and interconnect while increasing the die area. The greater number of transistors per integrated circuit with each new generation of process technology offers greater opportunities for enhanced circuit performance and functionality. The propagation delay through the individual circuit elements is reduced as the physical dimensions are scaled. Technology scaling-related enhancements coupled with advances in circuit structures and microarchitectures have significantly increased the performance of integrated circuits. The price for these performance and functional enhancements has traditionally been increased design complexity and higher power consumption.

Batteries provide the energy required to operate portable devices. Battery technologies have evolved at a much slower pace as compared to integrated circuit technologies. Due to the lack of a low cost battery technology with higher energy density characteristics as compared to the lithium-ion battery technology, enhancing performance, and functionality in portable devices has become increasingly challenging as increases in power consumption occur with each new technology generation. Another important consequence of the increasing power consumption is the degradation in the voltage quality and reliability of the power distribution networks. Due to scaling of the supply voltages coupled with higher power consumption, supply currents have increased significantly, producing metal migration and voltage fluctuations in the power distribution networks. Another important consequence of the increasing power consumption is the greater die temperature gradients and the formation of local hot spots which degrade the reliability and performance of high performance integrated circuits. As the power is consumed to produce a logic function, the energy is released in the form of heat which needs to be appropriately removed in order to maintain low ambient die temperatures. As the power density increases with greater power consumption and higher device and interconnect densities, it has become increasingly challenging to maintain a low ambient die temperature using traditional low cost cooling techniques based on heat sinks and air flow fans.

The generation, distribution, and dissipation of power are, therefore, now at the forefront in the design of high performance integrated circuits. Several techniques for designing low power and high speed integrated circuits are introduced in this book. Techniques for supply and threshold voltage scaling that lower power consumption or enhance device reliability without degrading circuit speed are described.

The dominant component of power consumption in CMOS circuits is dynamic switching power. Dynamic switching power is more than quadratically reduced by lowering the supply voltage. The other significant components of power consumption in current CMOS technologies such as the short-circuit, subthreshold leakage, and gate oxide leakage power are also significantly reduced at lower supply voltages. Scaling the supply voltage, therefore, is an effective strategy for lowering the power consumed by a CMOS circuit. As the supply voltage is reduced, however, the circuit speed degrades due to lower transistor currents. Systems that utilize multiple supply voltages, by selectively scaling the supply voltages along non-critical paths, can lower the speed penalty incurred in order to reduce the power consumption.

A primary issue in multiple supply voltage CMOS circuits is generating the multiple voltages. The energy and area overhead of the additional power supplies in multiple supply voltage CMOS integrated circuits is an important concern. These additional DC–DC converters are required to consume a small amount of energy in order to increase the energy savings attained with multiple supply voltage CMOS circuit techniques. High efficiency DC–DC conversion techniques with small area and good output voltage regulation characteristics are presented in this book. Monolithic DC–DC conversion on the same die as the load provides several desirable aspects. Integrating both the active and passive devices of a DC–DC converter onto the same die as a microprocessor increases energy efficiency, enhances the quality of the voltage regulation, and decreases the number of I/O pads dedicated for power delivery. Furthermore, by monolithically integrating the power supply, the reliability of the voltage conversion circuitry can be enhanced, the physical area can be reduced, and the overall cost of the DC–DC converter can be decreased as compared to a traditional discrete DC–DC converter.

An analysis of the power characteristics of a standard switching DC–DC converter topology, a buck converter, is presented. A parasitic circuit model is described for determining the optimum circuit configuration that produces the maximum efficiency. With this model, a closed form expression for the total power consumption of a buck converter is described. An analysis of an on-chip DC–DC converter over a wide range of design parameters is evaluated, permitting the development of a design space for fully integrating the active and passive devices of a DC–DC converter onto the same die. Full integration of a high efficiency buck converter on the same die as a dual supply voltage microprocessor is demonstrated to be feasible.

Two major challenges for a monolithic switching DC–DC converter are the area occupied by the integrated filter capacitor and the effect of the parasitic impedance characteristics of the integrated inductor on the overall efficiency characteristics of a switching DC–DC converter. A high switching frequency is the key design parameter that enables the integration of a high efficiency buck converter on the same die as a dual supply voltage microprocessor. An optimum switching frequency and inductor current ripple pair that maximizes the efficiency of a buck converter is shown to exist for a target technology. The global maximum efficiency is 92% at a switching frequency of 114 MHz for a voltage conversion from 1.2 V to 0.9 V while supplying 9.5 A of DC current, assuming an 80 nm CMOS technology. The required filter capacitance and inductance at this operating point are 2083 nF and 104 pH, respectively.

An efficiency of 88.4% is demonstrated at a switching frequency of 477 MHz when the filter capacitance is reduced to 100 nF due to tight area constraints on a microprocessor die. The area occupied by the buck converter is 12.6 mm^2 and is dominated by the area of the integrated filter capacitor. The analytic model for the converter efficiency is within 2.4% of the simulation results at the target design point.

For those high switching frequencies at which monolithic integration becomes feasible, the energy dissipated by the power MOSFETs and gate drivers dominates the total losses of the DC–DC converter. The efficiency can, therefore, be improved by applying a variety of MOSFET power reduction techniques. A low swing MOSFET gate drive technique is presented in this book that improves the efficiency of a DC–DC converter. A circuit model for low swing circuit optimization is also presented. The gate voltages and driver transistor sizes are included as independent parameters in the converter model. The optimum gate voltage swing of a power MOSFET that maximizes efficiency is shown to be lower than a standard full voltage swing. Lowering the voltage swing of the power MOSFET gate drivers is demonstrated to be effective for enhancing the efficiency characteristics of a DC–DC converter.

High voltage power delivery on a circuit board and monolithic DC–DC conversion provides enhanced voltage regulation quality and higher energy efficiency in the power generation and distribution network. Next generation low voltage and high power microprocessors are likely to require high input voltage, large step-down DC–DC converters monolithically integrated onto the same die. The voltage conversion ratios attainable with standard non-isolated switching DC–DC converter circuits are limited due to MOSFET reliability issues. Provided that a DC–DC converter is integrated onto the same die as a microprocessor (fabricated in a low voltage nanometer CMOS technology), the range of input voltages that can be applied to a standard DC–DC converter circuit are reduced. A standard non-isolated switching DC–DC converter topology such as a buck converter circuit is therefore not suitable for future high performance integrated circuits. High efficiency monolithic switching DC–DC converters that can generate very low operating voltages from a significantly higher board-level distribution voltage in a scaled nanometer CMOS technology are highly desirable.

Three step-down DC–DC converter topologies based on cascode buffers are introduced in this book for integration in a low voltage CMOS process. The cascode bridge circuits ensure that the voltages across the terminals of all of the MOSFETs in a DC–DC converter are maintained within the limits imposed by a standard low voltage CMOS technology. Reliable operation of the DC–DC converters operating at an input supply voltage up to three times as high as the maximum voltage that can be directly applied across the terminals of a MOSFET is verified assuming a 0.18 μm CMOS technology. The energy overhead of this circuit technique is low due to a charge recycling mechanism in the MOSFET gate drivers. An efficiency of 79.6% is demonstrated for a voltage conversion from 5.4 V to 0.9 V while supplying 250 mA of DC current.

Another important issue in multiple supply voltage circuits is high speed and full voltage swing signal transfer among the different voltage domains. Signal transfer among the circuitry operating at different voltage levels requires specialized voltage interface circuits. The power and delay overhead of voltage interface circuits is a primary issue in multiple supply voltage CMOS circuits. A bidirectional CMOS voltage interface circuit that drives high capacitive loads to full swing at high speed while consuming no static DC power is presented in this book. The propagation delay, power consumption, and power efficiency characteristics of the voltage interface circuit are compared to other interface circuits described in the literature. Up to a 3.6 times delay improvement and up to a 95% power reduction are observed as compared to previously published schemes. The speed and power characteristics of the voltage interface circuit have also been verified by experimental test circuits.

An alternative technique for reducing the impact of supply voltage scaling on circuit performance is scaling the threshold voltages. Threshold voltage scaling has accelerated during the past decade together with scaling the supply voltages. At reduced threshold voltages, subthreshold leakage currents increase exponentially. Supply voltage scaling when coupled with threshold voltage reduction, therefore, increase the leakage power while lowering the dynamic switching power. Similar to standard single supply voltage CMOS circuits, single threshold voltage CMOS circuits also suffer from excessive energy consumption in order to achieve a target throughput. The standard approach based on scaling the threshold voltage of an entire circuit so as to achieve a target signal propagation speed along a small number of critical delay paths is an inefficient method for enhancing performance.

A dynamic threshold voltage scaling technique can mitigate the deleterious side effects of threshold voltage scaling. Variable threshold voltage CMOS circuit techniques utilize the body terminal of the transistors to dynamically adjust the transistor currents during circuit operation. Dynamic threshold voltage scaling is typically used to reduce subthreshold leakage current produced by the idle portions of an integrated circuit while enhancing the speed of the active circuitry.

An application of variable threshold voltage CMOS circuit techniques is described in this book for enhanced noise immunity in domino logic circuits. A body bias circuit technique that provides significant enhancements in several electrical characteristics of domino logic circuits is presented. Standard domino logic circuits will become impractical within a few technology generations due to noise immunity issues. The variable threshold voltage keeper circuit technique can extend the life of domino logic circuits in high performance integrated circuits beyond the 45 nm CMOS technology generation. The technique is based on selectively applying reverse and forward body bias circuit techniques to a domino logic circuit. The circuit technique enhances the noise immunity of domino logic circuits without degrading the speed and power characteristics. By adjusting the current strength of a keeper transistor through a body bias, the electrical characteristics of a domino logic circuit are dynamically shifted toward either high speed/low power operation or higher noise immunity. Due to the dynamic nature of this technique (which automatically adjusts the keeper strength with respect to the changing speed, power, and noise immunity requirements of a domino logic circuit for different operational modes), the noise immunity, power, and speed characteristics can all be simultaneously enhanced.

Another technique for mitigating the deleterious effects of threshold voltage scaling is multiple threshold voltage CMOS. Multiple threshold voltage CMOS circuits offer decreased subthreshold leakage currents and enhanced performance by selectively lowering the threshold voltages along speed-critical paths.

A specific multiple threshold voltage CMOS circuit technique is presented in this book for application to high speed dynamic circuits. A quantitative study of the subthreshold leakage current characteristics of standard low threshold voltage and dual threshold voltage domino logic circuits is provided. Different subthreshold leakage current conduction paths exist with different dynamic and output node voltage states. A discharged dynamic node is shown to be preferable for reducing leakage current in a dual-V_t circuit. Alternatively, a charged dynamic node is preferred for lower subthreshold leakage energy in a standard low-V_t domino logic circuit with stacked pull-down devices, such as an AND gate.

Provided that a dual-V_t CMOS technology is employed, the noise immunity of domino logic circuits can be significantly degraded, affecting reliability. Two different dual-V_t domino logic circuit techniques can be employed in order to maintain similar noise immunity as compared to standard low-V_t circuits. Both keeper and output inverter sizing is required in a dual-V_t domino logic circuit with a high threshold voltage (high-V_t) keeper transistor in order to provide similar noise immunity as compared to a standard low-V_t domino logic circuit. An alternative dual-V_t domino logic circuit technique utilizes a low-V_t keeper transistor for enhanced noise immunity. Under similar noise immunity conditions as compared to standard low-V_t domino logic circuits, the savings in subthreshold leakage energy achieved by the dual-V_t circuit technique with a high-V_t keeper is 5.7 to 10.9 times higher as compared to the savings offered by the dual-V_t circuit technique with a low-V_t keeper. The effectiveness of the dual threshold voltage domino logic circuit technique in providing significant savings in subthreshold leakage energy is verified down to a 100 mV difference between the high and low threshold voltages in a 0.18 μm CMOS technology.

A circuit technique using sleep switch dual threshold voltage domino logic exploits the dynamic node voltage dependent asymmetry of the subthreshold leakage current characteristics of domino logic circuits. Existing techniques based on the application of a selected input vector to place a dual threshold voltage circuit into a low leakage state are shown to be ineffective in minimizing subthreshold leakage current in multiple stage domino circuits with inverted internal signals. The sleep switch circuit technique exploits the full effectiveness of a dual threshold voltage CMOS technology to reduce subthreshold leakage current by strongly turning off all of the high threshold voltage transistors, independent of the input signals. The energy overhead of the circuit technique is low, justifying the use of the sleep scheme by providing significant savings in total energy consumption during short idle periods.

Certain challenges in reducing power consumption while enhancing speed and maintaining reliability in CMOS integrated circuits have been highlighted in this book. Several enabling technologies have been described for achieving higher energy efficiency and enhanced reliability at the circuit and system levels. Examination of current integrated systems reveals inefficiencies at all levels of the design hierarchy. The complexity of the integrated systems is imposed by the production, distribution, and consumption dynamics of the marketplace. The enormous complexity of current integrated circuits composed of hundreds of millions of transistors is shaped by the market demand for ever increasing performance and functionality. Under ideal circumstances, attempts to cope with this system complexity would require frequent departures from traditional design methodologies. In reality, however, time-to-market constraints under the organizational hierarchy of the semiconductor industry dictate backward compatibility rather than breakthroughs. This market pressure has traditionally left little opportunity for significant innovation and optimization with each new technology generation. The quantitatively impressive enhancements in speed and variety of applications of integrated circuits over the years have historically been achieved by ignoring the increasing cost of energy and the degradation in reliability with each new product generation. While the additional market value created by enhanced speed (both clock speed and design turnaround) and functionality has been the primary focus, additional costs due to increasing power consumption and complexity have typically been neglected due to steadily increasing revenues. Continuing this same historical trend is not likely because currently the increasing cost imposed by greater levels of power consumption and degraded reliability will negate much of the additional value produced from the enhanced circuit speed and functionality.

Introducing a new product which satisfies both customer expectations for higher performance and greater functionality and vendor expectations for higher revenues will soon become infeasible based on existing company marketing policies which focus on the notion of higher speed at all costs.

The shift from more mainstream design approaches to more advanced design methodologies, as demonstrated throughout the history of the semiconductor industry, is imposed by the ever increasing requirements for enhanced productivity and cost effectiveness. A design methodology which provides a solution to a recognized problem can only survive in a market environment if the approach is cost effective. Due to lagging battery technologies, increasing cost of cooling, and decreasing yield (caused by the degradation of device, circuit, and system-level reliability), the authors of this book believe that the end of the road for current mainstream speed-centric CMOS design techniques is quickly approaching. Low power, signal integrity, and reliability concerns will dominate at all levels of the design hierarchy as the end of this speed-centric road that has been traveled for approximately four decades approaches. Low power and reliable integrated circuit and system design will develop into an increasingly exciting field, full of opportunities. The ideas and concepts presented in this book can be considered as a prelude to a larger discussion of the many opportunities for moving the performance and functionality of nanometer semiconductor technologies to even higher levels while remaining within a reasonable envelope of power consumption while providing satisfactory reliability.

Bibliography

1. Bohr MT. Nanotechnology goals and challenges for electronic applications. *IEEE Transactions on Nanotechnology.* 2002 March; 1 (1): 56–62.
2. Ronen R. *et al.* Coming challenges in microarchitecture and architecture. *Proceedings of the IEEE.* 2001 March; 89 (3): 325–340.
3. Borkar S. Design challenges of technology scaling. *IEEE Micro.* 1999 July/August; 19, 23–29.
4. Roy K, Prasad SC. *Low-Power CMOS VLSI Circuit Design.* John Wiley & Sons, Inc. 2000.
5. Borkar S. (September 2000). Obeying Moore's law beyond 0.18 micron. *Proceedings of the IEEE International ASIC/SOC Conference*, pp. 26–31.
6. Brooks DM. *et al.* Power-aware microarchitecture: design and modeling challenges for next generation microprocessors. *IEEE Micro.* 2000 November/December; 20, 26–44.
7. Flynn MJ, Hung P, Rudd KW. Deep submicron microprocessor design issues. *IEEE Micro.* 1999 July/August; 19, 11–22.
8. Takahashi O, Dhong SH, Hofstee P, Silberman J. (December 2001). High-Speed, power-conscious circuit design techniques for high-performance computing. *Proceedings of the IEEE International Symposium on VLSI Technology, Systems, and Applications*, pp. 279–282.
9. Chandrakasan AP, Brodersen RW. *Low Power CMOS Digital Design.* Kluwer Academic: Norwell, MA, 1995.
10. Gunther SH, Binns F, Carmean DM, Hall JC. Managing the impact of increasing microprocessor power consumption. *Intel Technology Journal*, 2001 February, Q1, 1–9.
11. Slawsby A. (June 2002). Taking Charge: Trends in Mobile Device Power Consumption. *Intel Corporation Internal Documents and Presentations* #27514, pp. 1–13.
12. Chandrakasan AP, Sheng S, Brodersen RW. Low-power CMOS digital design. *IEEE Journal of Solid-State Circuits.* 1992 April; 27 (4): 473–484.
13. Moore G. Cramming more components onto integrated circuits. *Electronics.* 1965 April; 38 (8): 114–117.
14. Moore GE. (December 1975). Progress in digital integrated electronics. *Proceedings of the IEEE International Electron Devices Meeting*, pp. 11–13.
15. Leblebici Y. Design considerations for CMOS digital circuits with improved hot-carrier reliability. *IEEE Journal of Solid-State Circuits.* 1996 July; 31 (7): 1014–1024.
16. Gelsinger PP, Gargini PA, Parker GH, Yu AYC. Microprocessors circa 2000. *IEEE Spectrum*, 1989 October, 43–47.
17. Sery G, Borkar S, De V. (June 2002). Life is CMOS: why chase the life after? *Proceedings of the IEEE/ACM Design Automation Conference*, pp. 78–83.

18. Pfiester JR, Shott JD, Meindl JD. Performance limits of CMOS ULSI. *IEEE Journal of Solid-State Circuits*, 1995 February; SC-20 (1): 253–263.

19. Chang L. *et al.* Moore's law lives on. *IEEE Circuits and Devices Magazine.* 2003 January; 19 (1): 35–42.

20. Klein T. Technology and performance of integrated complementary MOS circuits. *IEEE Journal of Solid-State Circuits*, 1969 June; SC-4 (3): 122–130.

21. Borkar S. (June 2001). Low power design challenges for the decade. *Proceedings of the IEEE/ACM Asia and South Pacific Design Automation Conference*, pp. 293–296.

22. *Intel Pentium 4 Processor Thermal Design Guide.* Intel Corporation Press, 2002.

23. Liu D, Svensson C. Trading speed for low power by choice of supply and threshold voltages. *IEEE Journal of Solid-State Circuits.* 1993 January; 28 (1): 10–17.

24. Song WS, Glasser LA. Power distribution techniques for VLSI circuits. *IEEE Journal of Solid-State Circuits*, 1986 February; SC-21 (1): 150–156.

25. Mezhiba AV, Friedman EG. Inductive properties of high-performance power distribution grids. *IEEE Transactions of Very Large Scale Integration (VLSI) Systems.* 2002 December; 10 (6): 762–776.

26. Kursun V, Narendra SG, De VK, Friedman EG. (September 2002). Efficiency analysis of a high frequency buck converter for on-chip integration with a dual-V_{DD} microprocessor. *Proceedings of the IEEE European Solid-State Circuits Conference*, pp. 743–746.

27. Gonzales R, Gordon BM, Horowitz MA. Supply and threshold voltage scaling for low power CMOS. *IEEE Journal of Solid-State Circuits.* 1997 August; 32 (8): 1210–1216.

28. Usami K. *et al.* Automated low-power technique exploiting multiple supply voltages applied to a media processor. *IEEE Journal of Solid-State Circuits.* 1998 March; 33 (3): 463–472.

29. Mutoh S. *et al.* 1-V power supply high-speed digital circuit technology with multithreshold-voltage CMOS. *IEEE Journal of Solid-State Circuits.* 1995 August; 30 (8): 847–854.

30. Kursun V, Narendra SG, De VK, Friedman EG. Analysis of buck converters for on-chip integration with a dual supply voltage microprocessor. *IEEE Transactions on Very Large Scale Integration (VLSI) Systems.* 2003 June; 11 (3): 514–522.

31. Kursun V, Narendra SG, De VK, Friedman EG. Monolithic (March 2003). Monolithic DC-DC converter analysis and MOSFET gate voltage optimization. *Proceedings of the IEEE International Symposium on Quality Electronic Design*, pp. 279–284.

32. Kursun V, Secareanu RM, Friedman EG. (May 2002). CMOS voltage interface circuit for low power systems. *Proceedings of the IEEE International Symposium on Circuits and Systems*, Vol. 3, pp. 667–670.

33. Kursun V, Friedman EG. Domino logic with variable threshold voltage keeper. *IEEE Transactions on Very Large Scale Integration (VLSI) Systems.* 2003 December; 11 (6): 1080–1093.

34. Kursun V, Friedman EG. (April 2002). Low swing dual threshold voltage domino logic. *Proceedings of the ACM/SIGDA Great Lakes Symposium on VLSI*, pp. 47–52.

35. Dropsho S, Kursun V, Albonesi DH, Dwarkadas S, Friedman EG. (November 2002). Managing static leakage energy in microprocessor functional units. *Proceedings of the IEEE/ACM International Symposium on Microarchitecture*, pp. 321–332.

36. Chandrakasan AP, Brodersen RW. Minimizing power consumption in digital CMOS circuits. *Proceedings of the IEEE.* 1995 April; 83 (4): 498–523.

37. Chandrakasan A, Bowhill WJ, Fox F. *Design of High-Performance Microprocessor Circuits.* New York: IEEE Press, 2001.

38. Nilsson JW. *Electric Circuits.* Addison-Wesley: Reading, MA, 1994.

39. Liu W. *et al. BSIM4.2.0 MOSFET Model Users' Manual.* University of California: Berkeley, 2000.

40. Frank DJ. *et al.* Device scaling limits of Si MOSFETs and their application dependencies. *Proceedings of the IEEE.* 2001 March; 89 (2): 259–288.

41. Taur Y, Wann CH, Frank DJ. (December 1998). 25 nm CMOS Design Considerations. *Proceedings of the IEEE International Electron Devices Meeting*, pp. 789–792.

42. Taur Y. (June 1999). CMOS Scaling Beyond 0.1 µm: How far can it go? *Proceedings of the IEEE International Symposium on VLSI Technology, Systems, and Applications*, pp. 6–9.

43. Frank DJ, Taur Y, Wong H-SP. (June 1999). Future prospects for Si CMOS technology. *Proceedings of the IEEE Annual Device Research Conference*, pp. 18–21.

44. Grotjohn T, Hoefflinger B. A parametric short-channel MOS transistor model for subthreshold and strong inversion current. *IEEE Journal of Solid-State Circuits*, 1984 February; SC-19 (1): 100–112.

45. Toyabe T, Asai S. Analytical models of threshold voltage and breakdown voltage of short-channel MOSFET's derived from two-dimensional analysis. *IEEE Journal of Solid-State Circuits*, 1979 April; SC-14 (2): 375–383.

46. Lin Y-S. *et al.* Leakage scaling in deep submicron CMOS for SoC. *IEEE Transactions on Electron Devices*. 2002 June; 49 (6): 1034–1041.

47. Zhang W-L, Tian L-L, Yang Z-L. (October 1998). Unified deep-submicron MOSFET model for circuit simulation. *Proceedings of the IEEE International Conference on Solid-State and Integrated Circuit Technology*, pp. 439–442.

48. Ferre A, Figueras J. (September 1998). Characterization of leakage power in CMOS technologies. *Proceedings of the IEEE International Conference on Electronics, Circuits and Systems*, Vol. 2, pp. 185–188.

49. Sheu BJ, Scharfetter DL, Ko P-K, Jeng M-C. BSIM: Berkeley short-channel IGFET model for MOS transistors. *IEEE Journal of Solid-State Circuits*, 1987 August; SC-22 (4): 558–566.

50. Narendra S, De V, Borkar S, Antoniadis D, Chandrakasan A. (August 2002). Full-chip Sub-threshold leakage power prediction model for sub-0.18 µm CMOS. *Proceedings of the IEEE International Symposium on Low Power Electronics and Design*, pp. 19–23.

51. Ghani T. *et al.* (December 1999). 100 nm gate length high performance/low power CMOS transistor structure. *Proceedings of the IEEE International Electron Devices Meeting*, pp. 415–418.

52. Thompson S. *et al.* (December 2001). An enhanced 130 nm generation logic technology featuring 60 nm transistors optimized for high performance and low power at 0.7–1.4 V. *Proceedings of the IEEE International Electron Devices Meeting*, pp. 257–260.

53. Cai J. *et al.* (June 2002). Supply voltage strategies for minimizing the power of CMOS processors. *Proceedings of the IEEE International Symposium on VLSI Technology*, pp. 102–103.

54. Thompson S. *et al.* (December 2002). An 90 nm logic technology featuring 50 nm strained silicon channel transistors, 7 layers of Cu interconnects, low k ILD, and 1 µm^2 SRAM cell. *Proceedings of the IEEE International Electron Devices Meeting*, pp. 61–64.

55. Ghani T. *et al.* (June 2000). Scaling challenges and device design requirements for high performance sub-50 nm gate length planar CMOS transistors. *Proceedings of the IEEE International Symposium on VLSI Technology*, pp. 174–175.

56. Plummer JD, Griffin PB. Material and process limits in silicon VLSI technology. *Proceedings of the IEEE*. 2001 March; 89 (3): 240–258.

57. Cao KM. *et al.* (December 2000). BSIM4 gate leakage model including source-drain partition. *Proceedings of the IEEE International Electron Devices Meeting*, pp. 815–818.

58. Linder BP. *et al.* Voltage dependence of hard breakdown growth and the reliability implication in thin dielectrics. *IEEE Electron Device Letters*. 2002 November; 23 (11): 661–663.

59. Linder BP. *et al.* (March–April 2003) Growth and scaling of oxide conduction after breakdown. *Proceedings of the IEEE International Reliability Physics Symposium*, pp. 402–405.

60. Mohapatra NR, Desai MP, Narendra SG, Rao VR. The effect of high-K gate dielectrics on deep submicrometer CMOS device and circuit performance. *IEEE Transaction on Electron Devices*. 2002 May; 49 (5): 826–831.

61. Veendrick HJM. Short-circuit dissipation of static CMOS circuitry and its impact on the design of buffer circuits. *IEEE Journal of Solid-State Circuits*, 1984 August; SC-19 (4): 468–473.

62. Nose K, Sakurai T. Analysis and future trend of short-circuit power. *IEEE Transactions on Computer-Aided Design of Integrated Circuits and Systems*. 2000 September; 19 (9): 1023–1030.

63. Narendra S, Antoniadis D, De V. (August 1999). Impact of using adaptive body bias to compensate die-to-die V_t variation on within-die V_t variation. *Proceedings of the IEEE International Symposium on Low Power Electronics and Design*, pp. 229–232.

64. Bowman KA, Duvall SG, Meindl JD. Impact of die-to-die and within-die parameter fluctuations on the maximum clock frequency distribution for gigascale integration. *IEEE Journal of Solid-State Circuits.* 2002 February; 37 (2): 183–190.

65. Shimohigashi K, Seki K. Low-voltage ULSI design. *IEEE Journal of Solid-State Circuits.* 1993 April; 28 (4): 408–413.

66. Sun S-W, Tsui PGY. Limitation of CMOS supply-voltage scaling by MOSFET threshold-voltage variation. *IEEE Journal of Solid-State Circuits.* 1995 August; 30 (8): 947–949.

67. Chen K, Hu C. Performance and V_{dd} scaling in deep submicrometer CMOS. *IEEE Journal of Solid-State Circuits.* 1998 October; 33 (10): 1586–1589.

68. Burd TD, Brodersen RW. *Energy Efficient Microprocessor Design.* Kluwer Academic, 2002.

69. Nowka KJ. *et al.* A 32-bit PowerPC system-on-a-chip with support for dynamic voltage scaling and dynamic frequency scaling. *IEEE Journal of Solid-State Circuits.* 2002 November; 37 (11): 1441–1447.

70. Burd TD, Pering TA, Stratakos AJ, Brodersen RW. A dynamic voltage scaled microprocessor system. *IEEE Journal of Solid-State Circuits.* 2000 November; 35 (11): 1571–1580.

71. Burd TD, Brodersen RW. (July 2000). Design issues for dynamic voltage scaling. *Proceedings of the IEEE International Symposium on Low Power Electronics and Design*, pp. 9–14.

72. Kuroda T. *et al.* A 0.9-V, 150-MHz, 10-mW, 4 mm^2, 2-D discrete cosine transform core processor with variable threshold-voltage (VT) scheme. *IEEE Journal of Solid-State Circuits.* 1996 November; 31 (11): 1770–1779.

73. Miyazaki M, Ono G, Ishibashi K. (July 1998). A delay distribution squeezing scheme with speed-adaptive threshold-voltage CMOS (SA-Vt CMOS) for low voltage LSIs. *Proceedings of the IEEE International Symposium on Low Power Electronics and Design*, pp. 48–53.

74. Huang S-F. *et al.* (June 2001). Scalability and biasing strategy for CMOS with active well bias. *Proceedings of the IEEE International Symposium on VLSI Technology*, pp. 107–108.

75. Keshavarzi A. *et al.* (July 1999). Technology scaling behavior of optimum reverse body bias for standby leakage power reduction in CMOS IC's. *Proceedings of the IEEE International Symposium on Low Power Electronics and Design*, pp. 252–254.

76. Keshavarzi A. *et al.* (August 2001). Effectiveness of reverse body bias for leakage control in scaled dual-Vt CMOS ICs. *Proceedings of the IEEE International Symposium on Low Power Electronics and Design*, pp. 207–212.

77. Wann C. *et al.* (June 2000). CMOS with active well bias for low-power and RF/analog applications. *Proceedings of the IEEE International Symposium on VLSI Technology*, pp. 158–159.

78. Narendra S. *et al.* Forward body bias for microprocessors in 130-nm technology generation and beyond. *IEEE Journal of Solid-State Circuits.* 2003 May; 38 (5): 696–701.

79. Miyazaki M, Ono G, Ishibashi K. A 1.2-GIPS/W microprocessor using speed-adaptive threshold-voltage CMOS with forward bias. *IEEE Journal of Solid-State Circuits.* 2002 February; 37 (2): 210–217.

80. Tschanz J. *et al.* Adaptive body bias for reducing impacts of die-to-die and within-die parameter variations on microprocessor frequency and leakage. *IEEE Journal of Solid-State Circuits.* 2002 November; 37 (11): 1396–1402.

81. Nose K. *et al.* V_{TH}-hopping scheme to reduce subthreshold leakage for low-power processors. *IEEE Journal of Solid-State Circuits.* 2002 March; 37 (3): 413–419.

82. Kindert N, Sugii T, Tang S, Hu C. Dynamic threshold pass-transistor logic for improved delay at lower power supply voltages. *IEEE Journal of Solid-State Circuits.* 1999 January; 34 (1): 85–89.

83. Mutoh S. *et al.* A 1-V multithreshold-voltage CMOS digital signal processor for mobile phone application. *IEEE Journal of Solid-State Circuits.* 1996 November; 31 (11): 1795–1802.

84. Sakata T, Itoh K, Horiguchi M, Aoki M. Subthreshold-current reduction circuits for multi-gigabit DRAM's. *IEEE Journal of Solid-State Circuits.* 1994 July; 29 (7): 761–769.

85. Mutoh S, Shigematsu S, Gotoh Y, Konaka S. (January 1999). Design method of MTCMOS power switch for low-voltage high-speed LSIs. *Proceedings of the IEEE Asia and South Pacific Design Automation Conference,* pp. 113–116.

86. Kao JT, Chandrakasan A. Dual-threshold voltage techniques for low-power digital circuits. *IEEE Journal of Solid-State Circuits.* 2000 July; 35 (7): 1009–1018.

87. Eisele M, Berthold J, Schmitt-Landsiedel D, Mahnkopf R. (August 1996). The impact of intra-die device parameter variations on path delays and on the design for yield of low voltage digital circuits. *Proceedings of the IEEE International Symposium on Low Power Electronics and Design,* pp. 237–242.

88. Takashima D. *et al.* Standby/active mode logic for sub-1-V operating ULSI memory. *IEEE Journal of Solid-State Circuits.* 1994 April; 29 (4): 441–447.

89. Su L. *et al.* (June 1998). A High-Performance Sub-0.25 μm CMOS technology with multiple thresholds and copper interconnects. *Proceedings of the IEEE International Symposium on VLSI Technology,* pp. 18–19.

90. McPherson T. *et al.* (February 2000). 760 MHz G6 S/390 microprocessor exploiting multiple Vt and copper interconnects. *Proceedings of the IEEE International Solid-State Circuits Conference,* pp. 96–97.

91. Sakurai T, Newton AR. A simple MOSFET model for circuit analysis. *IEEE Transactions on Electron Devices.* 1991 April; 38 (4): 887–894.

92. Khellah MM, Elmasry MI. (June 1999). Power minimization of high-performance submicron CMOS circuits using a dual-V_{dd} Dual-V_{th} (DVDV) approach. *Proceedings of the IEEE International Symposium on Low Power Electronics and Design,* pp. 106–108.

93. Kuroda T. *et al.* Variable supply-voltage scheme for low-power high-speed CMOS digital design. *IEEE Journal of Solid-State Circuits.* 1998 March; 33 (3): 454–462.

94. Takahashi M. *et al.* A 60-mW MPEG4 Video Codec using clustered voltage scaling with variable supply-voltage scheme. *IEEE Journal of Solid-State Circuits.* 1998 November; 33 (11): 1772–1780.

95. Kuroda T, Hamada M. Low-power CMOS digital design with dual embedded adaptive power supplies. *IEEE Journal of Solid-State Circuits.* 2000 April; 35 (4): 652–655.

96. Hamada M. *et al.* (May 1998). A top-down low power design technique using clustered voltage scaling with variable supply-voltage scheme. *Proceedings of the IEEE Custom Integrated Circuits Conference,* pp. 495–498.

97. Kao JT, Miyazaki M, Chandrakasan AP. A 175-mV multiply-accumulate unit using an adaptive supply voltage and body bias architecture. *IEEE Journal of Solid-State Circuits.* 2002 November; 37 (11): 1545–1554.

98. Tschanz J, Narendra S, Nair R, De V. Effectiveness of adaptive supply voltage and body bias for reducing impact of parameter variations in low power and high performance microprocessors. *IEEE Journal of Solid-State Circuits.* 2003 May; 38 (5): 826–829.

99. Soeleman H, Roy K, Paul BC. Robust subthreshold logic for ultra-low power operation. *IEEE Transactions on Very Large Scale Integration (VLSI) Systems.* 2001 February; 9 (1): 90–99.

100. Tsividis Y. *Operation Modeling of the MOS Transistor.* The McGraw-Hill: New York, 1999.

101. Panov Y, Jovanovic MM. Design and performance evaluation of low-voltage/high-current DC/DC on-board modules. *IEEE Transactions on Power Electronics.* 2001 January; 16 (1): 26–33.

102. Erickson RW, Maksimovic D. *Fundamentals of Power Electronics.* Kluwer Academic: Norwell, MA, 2001.

103. Furuyama T, Watanabe Y, Ohsawa T, Watanabe S. A new on-chip voltage converter for submicrometer high-density DRAM's. *IEEE Journal of Solid-State Circuits,* 1987 June; SC-22 (3): 437–440.

104. Takashima D. *et al.* Low-power on-chip supply voltage conversion scheme for ultrahigh-density DRAM's. *IEEE Journal of Solid-State Circuits.* 1993 April; 28 (4): 504–509.

105. Ooishi T. *et al.* A mixed-mode voltage down converter with impedance adjustment circuitry for low-voltage high-frequency memories. *IEEE Journal of Solid-State Circuits.* 1996 April; 31 (4): 575–585.

106. Endoh T, Sunaga K, Sakuraba H, Masuoka F. An on-chip 96.5% current efficiency CMOS linear regulator using a flexible control technique of output current. *IEEE Journal of Solid-State Circuits.* 2001 January; 36 (1): 34–38.

107. Wang C-C, Wu J-C. Efficiency improvement in charge pump circuits. *IEEE Journal of Solid-State Circuits.* 1997 June; 32 (6): 852–860.

108. Arntzen B, Maksimovic D. Switched-capacitor DC/DC converters with resonant gate drive. *IEEE Transactions on Power Electronics.* 1998 September; 13 (5): 892–902.

109. Maksimovic D, Dhar S. (June 1999). Switched-capacitor DC–DC converters for low-power on-chip applications. *Proceedings of the IEEE Power Electronics Specialists Conference,* pp. 54–59.

110. Blanchard R, Thibodeau PE. (June 1985). The design of a high efficiency, low voltage power supply using MOSFET synchronous rectification and current mode control. *Proceedings of the IEEE Power Electronics Specialists Conference,* pp. 355–361.

111. Kagan RS, Chi M. (July 1982). Improving power supply efficiency with MOSFET synchronous rectifiers. *Proceedings of the International Solid-State Power Conversion Conference,* pp. D4.1–4.9.

112. Reynolds SK. A DC-DC converter for short-channel CMOS technologies. *IEEE Journal of Solid-State Circuits.* 1997 January; 32 (1): 111–113.

113. Stratakos A, Sanders SR, Brodersen RW. (April 1994). A low-voltage CMOS DC–DC converter for a portable battery-operated system. *Proceedings of the IEEE Power Electronics Specialists Conference,* pp. 619–626.

114. Arbetter B, Maksimovic D. (April 1998). DC–DC converter with fast transient response and high efficiency for low-voltage microprocessor loads. *Proceedings of the IEEE Applied Power Electronics Conference,* pp. 156–162.

115. Arbetter B, Maksimovic D. (April 1997). Control method for low-voltage DC power supply in battery-powered systems with power management. *Proceedings of the IEEE Power Electronics Specialists Conference,* pp. 1198–1204.

116. Weinberg SH. (1992). A novel lossless resonant MOSFET driver. *Proceedings of the IEEE Power Electronics Specialists Conference,* pp. 1003–1010.

117. Maksimovic D. (April 1991). A MOS gate drive with resonant transitions. *Proceedings of the IEEE Power Electronics Specialists Conference,* pp. 527–532.

118. Gronowski PE. *et al.* High-performance microprocessor design. *IEEE Journal of Solid-State Circuits.* 1998 May; 33 (5): 676–686.

119. Gardner D, Crawford AM, Wang S. (June 2001). High frequency (GHz) and low resistance integrated inductors using magnetic materials. *Proceedings of the IEEE International Interconnect Technology Conference,* pp. 101–103.

120. Cherkauer BS, Friedman EG. A unified design methodology for CMOS tapered buffers. *IEEE Transactions on Very Large Scale Integration (VLSI) Systems.* 1995 March; 3 (1): 99–111.

121. Gardner D. (2001) Personal Communication. Intel Corporation, Components Research, Santa Clara, CA.

122. Secareanu RM, Friedman EG. (May 1999). A universal CMOS voltage interface circuit. *Proceedings of the IEEE International Symposium on Circuits and Systems,* pp. 1242–1245.

123. Caravella JS, Quigley JH. (September 1993). Three volt to five volt CMOS interface circuit with device leakage limited DC power dissipation. *Proceedings of the IEEE ASIC Conference,* pp. 448–451.

124. Golshan R, Haroun B. (June 1994). A novel reduced swing CMOS bus interface circuit for high speed low power VLSI systems. *Proceedings of the IEEE International Symposium on Circuits and Systems,* Vol. 4, pp. 351–354.

125. Zhang H, George V, Rabaey JM. Low-swing on-chip signaling techniques: effectiveness and robustness. *IEEE Transactions on VLSI Systems*. 2000 June; 8 (3): 264–272.

126. Nakagome Y. *et al.* Sub 1-V swing internal bus architecture for future low-power ULSI's. *IEEE Journal of Solid-State Circuits*. 1993 April; 28 (4): 414–419.

127. Nowka KJ, Galambos T. (October 1998). Circuit design techniques for a gigahertz integer microprocessor. *Proceedings of the IEEE International Conference on Computer Design*, pp. 11–16.

128. Alvandpour A, Larsson-Edefors P, Svensson C. (September 1999). A leakage tolerant multi-phase keeper for wide domino circuits. *Proceedings of the IEEE International Conference on Electronics, Circuits and Systems*, pp. 209–212.

129. Alvandpour A, Krishnamurty RK, Soumyanath K, Borkar SY. A sub-130-nm conditional keeper technique. *IEEE Journal of Solid-State Circuits*. 2002 May; 37 (5): 633–638.

130. Allam MW, Anis MH, Elmasry MI. (July 2000). High-speed dynamic logic styles for scaled-down CMOS and MTCMOS technologies. *Proceedings of the IEEE International Symposium on Low Power Electronics and Design*, pp. 155–160.

131. Keshavarzi A, Narendra S, Bloechel B, Borkar S, De V. (June 2002). Forward body bias for microprocessors in 130 nm technology generation and beyond. *Proceedings of the IEEE International Symposium on VLSI Circuits*, pp. 312–315.

132. Tschanz J, Narendra S, Nair R, De V. (June 2002). Effectiveness of adaptive supply voltage and body bias for reducing the impact of parameter variations in low power and high performance microprocessors. *Proceedings of the IEEE International Symposium on VLSI Circuits*, pp. 310–311.

133. Hwang IS, Fisher AL. Ultrafast compact 32-bit CMOS adders in multiple-output domino logic. *IEEE Journal of Solid-State Circuits*. 1989 April; 24 (2): 358–369.

134. Srivastava P, Pua A, Welch L. (February 1998). Issues in the design of domino logic circuits. *Proceedings of the IEEE Great Lakes Symposium on VLSI*, pp. 108–112.

135. Rusu S, Singer G. The first IA-64 microprocessor. *IEEE Journal of Solid-State Circuits*. 2000 November; 35 (11): 1539–1544.

136. Kao J. (September 1999). Dual threshold voltage domino logic. *Proceedings of the European Solid-State Circuits Conference*, pp. 118–121.

137. Silberman J. *et al.* A 1.0-GHz single-issue 64-bit PowerPC integer processor. *IEEE Journal of Solid-State Circuits*. 1998 November; 33 (11): 1600–1608.

138. Balamurugan G, Shanbhag NR. (August 1999). Energy-efficient dynamic circuit design in the presence of crosstalk noise. *Proceedings of the IEEE International Symposium on Low Power Electronics and Design*, pp. 24–29.

139. Rjoub A, Koufopavlou O, Nikolaidis S. (May 1998). Low-power/low swing domino CMOS logic. *Proceedings of the IEEE International Symposium on Circuits and Systems*, Vol. 2, pp. 13–16.

140. Shieh S, Wang J, Yeh Y. (September 2001). A contention-alleviated static keeper for high-performance domino logic circuits. *Proceedings of the IEEE International Conference on Electronics, Circuits, and Systems*, Vol. 2, pp. 707–710.

141. Jung S, Yoo S, Kim K, Kang S. (May 2001). Skew-tolerant high-speed (STHS) domino logic. *Proceedings of the IEEE International Symposium on Circuits and Systems*, Vol. 4, pp. 154–157.

142. Heo S, Asanovic K. (June 2002). Leakage-biased domino circuits for dynamic fine-grain leakage reduction. *Proceedings of the IEEE International Symposium on VLSI Circuits*, pp. 316–319.

143. Ye Y, Borkar S, De V. (June 1998). A new technique for standby leakage reduction in high-performance circuits. *Proceedings of the IEEE International Symposium on VLSI Circuits*, pp. 40–41.

144. Wang C-C, Lee P-M, Chen K-L. An SRAM design using dual threshold voltage transistors and low-power quenchers. *IEEE Journal of Solid-State Circuits*. 2003 October; 38 (10): 1712–1720.

145. Krishnamurthy RK, Alvandpour A, De V, Borkar S. (May 2002). High-performance and low-power challenges for sub-70 nm microprocessor circuits. *Proceedings of the IEEE Custom Integrated Circuits Conference*, pp. 125–128.

146. Narendra S. *et al.* (August 2001). Scaling of stack effect and its application for leakage reduction. *Proceedings of the IEEE/ACM International Symposium on Low Power Electronics and Design,* pp. 195–200.

147. Park J-T, Colinge J-P. Multiple-gate SOI MOSFETs: device design guidelines. *IEEE Transactions on Electron Devices.* 2002 December; 49 (12): 2222–2229.

148. Greve DW. *Field Effect Devices and Applications.* Prentice-Hall, Englewood Cliffs, NJ, 1998.

149. Schroder DK. Low power silicon devices. In: Buschow KHJ, Cahn RW, Flemings MC, Ilschner B, Kramer EJ, Mahajan S. *The Encyclopedia of Materials: Science and Technology.* Elsevier: Amsterdam, 2001.

150. Chuang CT, Lu PF, Anderson CJ. SOI for digital CMOS VLSI: design considerations and advances. *Proceedings of the IEEE.* 1998 April; 86 (4): 689–720.

151. Hammad MY, Schroder DK. Analytical modeling of the partially-depleted SOI MOSFET. *IEEE Transactions on Electron Devices.* 2001 February; 48 (2): 252–258.

152. Liu Y. *et al.* Systematic electrical characteristics of ideal rectangular cross section Si-fin channel double-gate MOSFETs fabricated by a wet process. *IEEE Transactions on Nanotechnology.* 2003 December; 2 (4): 198–204.

153. Xu J. Nanotube electronics: non-CMOS routes. *Proceedings of the IEEE.* 2003 November; 91 (11): 1819–1829.

154. Avouris P. *et al.* Carbon nanotube electronics. *Proceedings of the IEEE.* 2003 November; 91 (11): 1772–1784.

155. Magklis G. *et al.* Dynamic frequency and voltage scaling for a multiple-clock domain micro-processor. *IEEE Micro.* 2003 November/December; 23 (6): 62–68.

156. Semeraro G. *et al.* (February 2002). Energy-efficient processor design using multiple clock domains with dynamic voltage and frequency scaling. *Proceedings of the IEEE International Symposium on High-Performance Computer Architecture,* pp. 29–40.

157. Lee D, Blaauw D, Sylvester D. Gate oxide leakage current analysis and reduction for VLSI circuits. *IEEE Transactions on Very Large Scale Integration (VLSI) systems.* 2004 February; 12 (2): 155–166.

158. Kumar R. Interconnect and noise immunity design for the Pentium 4 processor. *Intel Technology Journal,* 2001; (Q1): 1–12.

159. Shepard KL, Narayanan V. Conquering noise in deep-submicron digital ICs. *IEEE Design and Test of Computers,* 1998 January–March: 15 (1): 51–62.

160. Tang KT, Friedman EG. Simultaneous switching noise in on-chip CMOS power distribution networks. *IEEE Transactions on Very Large Scale Integration (VLSI) Systems.* 2002 August; 10 (4): 487–493.

161. Grove A. (December 2002). Changing Vectors of Moore's Law. *International Electron Devices Meeting.*

162. 2001 International Technology Roadmap for Semiconductors, http: //public.itrs.net.

163. Ghani T. *et al.* (June 2000). Scaling challenges and device design requirements for high performance sub-50 nm gate length planar CMOS transistors. *Proceedings of the IEEE International Symposium on VLSI Technology,* pp. 174–175.

164. Mezhiba AV, Friedman EG. *Power Distribution Networks in High Speed Integrated Circuits.* Kluwer Academic Publishers: Norwell, MA, 2004.

165. Wanlass FM, Sah CT. (February 1963). Nanowatt logic using field-effect metal-oxide semiconductor triodes. *Proceedings of the IEEE International Solid-State Circuits Conference,* Vol. 6, pp. 32–33.

166. Kang S-M, Leblebici Y. *CMOS Digital Integrated Circuits.* McGraw-Hill: New York, 1999.

167. Liu Z, Kursun V. (August 2005). Temperature dependent leakage power characteristics of dynamic circuits in sub-65 nm CMOS technologies. *Proceedings of the IEEE International Midwest Symposium on Circuits and Systems.*

168. Liu Z, Kursun V. (September 2005). Shifted leakage power characteristics of dynamic circuits due to gate oxide tunneling. *Proceedings of the IEEE International Systems on Chip (SOC) Conference,* pp. 151–154.

169. Kursun V, Schrom G, De VK, Friedman EG, Narendra SG. (May 2005). Cascode buffer for monolithic voltage conversion operating at high input supply voltages. *Proceedings of the IEEE International Symposium on Circuits and Systems*, pp. 464–467.

170. Kursun V, De VK, Friedman EG, Narendra SG. Monolithic voltage conversion in low voltage CMOS technologies. *Microelectronics Journal*. 2005 September; 36 (9): 863–867.

171. Kursun V, Narendra SG, De VK, Friedman EG. Cascode monolithic DC-DC converter for reliable operation at high input voltages. *International Journal of Analog Integrated Circuits and Signal Processing*. 2005 March; 42 (3): 231–238.

172. Kursun V, Narendra SG, De VK, Friedman EG. Low voltage swing monolithic DC-DC conversion. *IEEE Transactions on Circuits and Systems II*. 2004 May; 51 (5): 241–248.

173. Kursun V, Friedman EG. Sleep switch dual threshold voltage domino logic with reduced standby leakage current. *IEEE Transactions on Very Large Scale Integration (VLSI) Systems*. 2004 May; 12 (5): 485–496.

174. Friedman EG. *Clock Distribution Networks in VLSI Circuits and Systems*. Piscataway, IEEE Press: Piscataway, NJ, 1995.

175. Lee W-C, Hu C. (June 2000). Modeling gate and substrate currents due to conduction- and valence-band electron and hole tunneling. *Proceedings of the IEEE International Symposium on VLSI Technology*, pp. 198–199.

176. Yeo YC. *et al.* Direct tunneling gate leakage current in transistors with ultrathin silicon nitride gate dielectric. *IEEE Electron Device Letters*. 2000 November; 21 (11): 540–542.

177. Lee W-C, Hu C. Modeling CMOS tunneling currents through ultrathin gate oxide due to conduction- and valence-band electron and hole tunneling. *IEEE Transactions on Electron Devices*. 2001 July; 48 (7): 1366–1373.

178. Schuegraf KF, King CC, Hu C. (June 1992). Ultra-thin silicon dioxide leakage current and scaling limit. *Proceedings of the IEEE International Symposium on VLSI Technology*, pp. 18–19.

179. Taur Y, Ning TH. *Fundamentals of Modern VLSI Devices*. Cambridge University Press: New york 2002.

Index

Printed and bound by CPI Group (UK) Ltd, Croydon, CR0 4YY

16/04/2025

14658461-0002